Mass Communication in Agricultural Extension

Mass Communication in Agricultural Extension

By :

D.K.Dangi
K.S. Kadian
&
O.P. Garhwal

SSPH

SATISH SERIAL PUBLISHING HOUSE
403, Express Tower, Commercial Complex, Azadpur, Delhi-110033 (India)
Phone : 011-27672852, Fax : 91-11-27672046
E-mail :info@satishserial.com, hkjain1975@yahoo.com
Website : www.satishserial.com

Published by :

SATISH SERIAL PUBLISHING HOUSE

403, Express Tower, Commercial Complex, Azadpur, Delhi-110033 (INDIA)
Phone : 011-27672852 Fax : 91-11-27672046
E-mail : info@satishserial.com, hkjain1975@yahoo.com

ISBN 978-93-81226-13-1

Reprint, 2023

Composed, Designed & Printed in India

Dedicated to

ALMITY

About the Authors

Dr. Dileep Kumar Dangi, born in Jhunjunu Rajasthan did his B.Sc. Agriculture (Hons.) and M.Sc. (Ag) Extension from S.K.R.A.U.Bikaner, Rajasthan, India, in 2001 and 2004 respectively. He topped the college and secured 3rd position at university level in Graduation. In the same university, he conducted the research on adoption of sprinkler system of irrigation in the Jhunjhunu District of Rajasthan. He did his Ph.D. Degree from N.D.R.I. Karnal, Haryana, India. In the same research Institute he had conducted research on *Critical analysis of Dairy Cooperatives and Private Dairies in Rajasthan*. He is closely associated with many national and international professional societies and authored many research and popular papers. He is also persuing MJMC (Master in journalism and mass communication) from IGNOU, Delhi.

Dr. K.S. Kadian, born in 1961 and did his B.Sc. (Hons.) Agri. from Haryana Agricultural University, Hisar in 1983. He did his M.Sc. and Ph. D. in the discipline of Dairy Extension Education at National Dairy Research Institute, (Deemed University) Karnal. Dr. Kadian started his career in 1989 as Assistant Professor in Agricultural Extension at Dr. Y. S. Parmar University of Horticulture & Forestry, Solan (Himachal Pradesh). He joined ARS as a Senior Scientist and posted at ICAR Research Complex for NEH Region, Barapani (Meghalaya) in 2000. Later on he was transferred from Barapani to National Dairy Research Institute, Karnal in 2002. He was promoted as a Principal Scientist from December, 2007. He has specialization in teaching courses like training methodology, Information & Communication Technologies and Advances in Extension. He has also supervised four M.Sc. and Six Ph.D. students. He was the recipient of Young Scientist Award conferred by Society of Extension Education. He is a life member of Indian Society of Extension Education, Society of Extension Education and Indian Dairy Association. He has authored two books, three teaching manuals and written more than sixty research papers published in Indian and foreign journals. At present, he is engaged in teaching, research and extension activities of Dairy Extension Division of N.D.R.I., Karnal.

Dr. O.P. Garhwal has obtained his B.Sc. (Ag), M.Sc. (Hort.) and Ph.D. (Hort.) from RAU, Bikaner (Presently SKRAU, Bikaner).He is a recipient of the prestigious University Gold Medal for his outstanding academic performance in Masters degree programme from SKRAU, Bikaner. Dr.Garhwal joined the University as Assistant Professor (Horticulture) in 2005. He distinguished himself as a Horticulturist as well as an Extension Scientist. He has more than 13 years of experience in field of Ag. Extension & published several Research papers, popular articles in various reputed journals/magazines etc. and led many outreach Programmes for the farming community like trainings, demonstrations, On Farm Trials & other village linkage activities.

Preface

We have great pleasure in presenting the first edition of a book on Mass Communication in Agricultural Extension. It is concise, comprehensive and complete.

This book has been written keeping in mind latest developments in the area of agricultural extension. It contains all the elements of mass media and its use in communication. The present work has been designed to equiped students and other curious readers with as much knowledge on all the topics of mass media communication as is desirable from point of view of brilliant success and understanding in the academic examinations.

The book covers all the topics of the revised syllabi of ICAR's approved Agricultural Extension, Veterinary Extension, Fishery Extension and Home Science Extension. The authors felt compelled to write a new book which provides basic concepts with explanation, review of recent literature and experiences, about mass communication.

We shall be grateful to the readers if they sent us their valuable and constructive comments so that further improvement may be made in the forthcoming editions of the book.

Authors

Contents

Chapter 1

Introduction to Mass Communication

Mass communication is the term used to describe the academic study of the various means by which individuals and entities relay information through mass media to large segments of the population at the same time. It is usually understood to relate to newspaper and magazine publishing, radio, television and film, as these are used both for disseminating news and for advertising.Mass communication research includes media institutions and processes such as diffusion of information, and media effects such as persuasion or manipulation of public opinion.

In addition to studying practical skills of journalism, public relations or advertising, they offer programs on "mass communication" or "mass communication research." The latter is often the title given to doctoral studies in such schools, whether the focus of the student's research is journalism practice, history, law or media effects. Departmental structures within such colleges may separate research and instruction in professional or technical aspects of mass communication. With the increased role of the Internet in delivering news and information, mass communication studies and media organizations tend to focus on the convergence of publishing, broadcasting and digital communication.

The academic mass communication discipline historically differs from media studies and communication studies programs with roots in departments of theatre, film or speech, and with more interest in "qualitative," interpretive theory, critical or cultural approaches to communication study. In contrast, many mass communication programs historically lean toward empirical analysis and quantitative research—from statistical content analysis of media messages to survey research, public opinion polling, and experimental research. Interest in "New Media" and "Computer Mediated Communication" is growing much faster than educational institutions can assimilate it. So far, traditional classes and degree programs have not been able to accommodate new shifts of the paradigm in communication technologies.[Although national standards for the study of interactive media have been present in the U.K. since the mid-nineties, course work in these areas tends to vary significantly from university to university. Graduates of Mass Communication programs work in a variety of fields in traditional news media and publishing, advertising, public relations and research institutes. Such programs are accredited by the Accrediting Council on Education in Journalism and Mass Communication (ACEJMC).

The Association for Education in Journalism and Mass CommunicationAEJMC.is the major membership organization for academics in the field, offering regional and national conferences and refereed publications. The International Communication AssociationICA.and National Communication Association (formerly the Speech Communication Association) include divisions and publications that overlap with those of AEJMC, but AEJMC historically has stronger ties to the mass communication professions in the United States.

The term 'mass' denotes great volume, range or extent (of people or production) and reception of messages.McQuail: McQuail's Mass Communication Theory. The important point about 'mass' is not that a given number of individuals receive the products, but rather that the products are available in principle to a plurality of recipients.Thompson: The Media and Modernity,

The term 'mass' suggests that the recipients of media products constitute a vast sea of passive, undifferentiated individuals. This is an image associated with some earlier critiques of 'mass culture' and Mass society which generally assumed that the development of mass communication has had a largely negative impact on modern social life, creating a kind of bland and homogeneous culture which entertains individuals without challenging them.Thompson: The Media and Modernity, 14. However, with the advancement in Media Technology, people are no longer receiving gratification without questioning the grounds on which it is based.Thompson: The Media and Modernity,.Instead, people are engaging themselves more with media products such as computers, cell phones and Internet.

These have gradually become vital tools for communications in society today. The aspect of 'communication' refers to the giving and taking of meaning, the transmission and reception of messages. The word 'communication' is really equated with 'transmission', as viewed by the sender, rather than in the fuller meaning, which includes the notions of response, sharing and interaction.McQuail: McQuail's Mass Communication Theory, Messages are produced by one set of individuals and transmitted to others who are typically situated in settings that are spatially and temporally remote from the original context of production. Therefore, the term 'communication' in this context masks the social and industrial nature of the media, promoting a tendency to think of them as interpersonal communication.Hartley: "Mass communication". Furthermore, it is known that recipients today do have some capacity to intervene in and contribute to the course and content of the communicative process.Thompson: The Media and Modernity, They are being both active and creative towards the messages that they are conveyed of. With the complement of the cyberspace supported by the Internet, not only those recipients are participants in a structured process of symbolic transmissionThompson: The Media and Modernity, p. constraints such as time and space are reordered and eliminated. 'Mass communication' can be seen as institutionalized production and generalized diffusion of symbolic goods via the fixation and transmission of information or symbolic content. It is known that the systems of information codification has shifted from analog to digital.Thompson: The Media and Modernity,.This has indeed advanced the communication between individuals. With the existence of Infrared, Bluetoothand Wi-Fi, cell phones are no longer solely a tool for audio transmission. We can transfer photos, music documents or even games and email at any time and anywhere. The development of media technology has indeed advanced the transmission rate and stability of information exchange.

Characteristics of Mass Communication: Five characteristics of mass communication have been identified by Cambridge University's John Thompson.

Firstly, it "comprises both technical and institutional methods of production and distribution"Thompson: The Media and Modernity, p. 15.. This is evident throughout the history of the media, from print to the Internet, each suitable for commercial utility.

Secondly, it involves the "commodification of symbolic forms", Thompson: The Media and Modernity, p. 16. as the production of materials relies on its ability to manufacture and sell large quantities of the work. Just as radio stations rely on its time sold to advertisements, newspapers rely for the same reasons on its space.

Mass communication's third characteristic is the "separate contexts between the production and reception of information", Thompson: The Media and Modernity, while the fourth is in its "reach to those 'far removed' in time and space, in comparison to the producers".Thompson: The Media and Modernity.

Mass communication, which involves "information distribution". This is a "one to many" form of communication, whereby products are mass produced and disseminated to a great quantity of audiences. Thompson: The Media and Modernity.

Mass Communication in India

The recent past has seen India awaken to the call of media and this fact has led to major developments in the Mass Communication study-sectors across India, which in turn has led to the establishment of various media houses- be it newspaper agencies or TV channels or advertising firms, etc. Today, the media plays a very important role in every individual's life.

Initially, common people were not very bothered about what the media was doing and media was strictly restricted to the government and the business houses. But today, every common man's problem is constantly reflected through media. And there is hardly any section of the society or any issues that are not taken up by the media. In fact, media has become a platform where common men can raise their problems and demand justice to any issue that is in circulation. Media has indeed become an integral part of a common man's life.

Agricultural communication refers to communicators aiming to talk about agriculture with both agricultural audiences and non-agricultural audiences. It is done formally and informally by agricultural extension and is considered a subset of science communication. However, it has evolved into its own professional field.

By definition, agricultural communicators are science communicators that deal exclusively with the diverse, applied science and business that is agriculture. An agricultural communicator is "expected to bring with him or her a level of specialized knowledge in the agricultural field that typically is not required of the mass communicator".

A growing market for agricultural journalists and broadcasters led to the establishment of agricultural journalism and agricultural communication academic disciplines.

The job market for agricultural communicators includes:

Farm broadcasting

Journalists and editors of agricultural magazines

Communication specialist, public relations practitioner, or Web developer for agricultural commodity organizations, businesses, non-profits

Sales representative for agricultural business

Science journalist

Land-grant university communication specialist

History

The academic field originated from communication courses that taught students in the agricultural sciences how to communicate. Originally, agricultural journalists were needed to report farm news for a much larger agricultural and rural audience. As people moved from the farm to cities and suburbs, a much greater proportion of the population had less direct knowledge and experience regarding agriculture. While a need still exists for agricultural journalists, an equal, if not greater need exists for agricultural communicators who can act as liaisons between an industry with deeply rooted traditions and values and a public with little to no understanding of how agriculture operates and why it is the way it is.

Research

The key journal in the field is the Journal of Applied Communications[3]. Researchers have focused on a variety of areas examining consumer attitudes toward agricultural products and practices including genetic engineering and genetically modified food[5], natural and organic food and production[, and food-related risks[7]. Another area of research has been media coverage of agriculture and agricultural issues. Topics have included media coverage of Bovine Spongiform Encephalopathy (mad cow disease), YouTube videos of California Proposition 2 (2008), and television news coverage of food safety scares[11].

The Agricultural Communications Documentation Center[12], maintained by the University of Illinois, compiles research and articles related to agriculture and communications as well.

Approaches to Agricultural Communication

Theoretically speaking, agricultural communication is an applied theoretical field. The academic curriculum and scholarly endeavors typically stay within the context of agriculture, natural resources, and occasionally, the life sciences. It examines communication and human dimension issues as they relate to a variety of issues in agriculture and natural resources. Agricultural journalism is not always differentiated from agricultural communications in research. One could argue that when research focuses on media coverage of agricultural issues or when it examines issues within agricultural journalism specifically (i.e., what influences editors of agriculture magazines to publish risk information), then it is more within the realm of agricultural journalism. Journalism is often seen as a subset of communication that is supposed to be fair and balanced like traditional journalism, whereas the broader field of agricultural communication could potentially be viewed as advocacy communication.

Agricultural communicators are expected to have a certain amount of knowledge and familiarity with agriculture. One could also add to that definition and say the communicator also brings with him or her appreciation, or even affection, for the agriculture industry. While this is also probably true of agricultural journalists, they at least need to be cognizant of their potential bias to ensure they ask critical questions and present unbiased information. Agricultural journalists are trained like traditional journalists, but bring with them an understanding of agricultural systems and science either through experience and/or academic training.

❑❑❑

Chapter 2

Basics and Importance of Mass Media

Introduction

Mass media - written, spoken and visual carry information from its source to receivers - readers, listeners, viewers. The media may be used effectively in dealing with all sections of society everywhere, but they are especially important for taking science and technology to Indian agriculture.

Mass media can be further divided into four types - printed, spoken, visual, and a combination of those three. The term mass media implies a large andience. Mass media materials are needed whenever large number of people must be reached and motivated to action, which is an important advantage. Radio, television, cinema and newspapers are examples of mass media which have large audiences with very low cost per person contacted. Where mass media material influence people to adopt new practices, the cost per adoption is low. These materials may be used to supplement nearly any other teaching and information method.

A practical extension teaching example will serve to illustrate this point. The average village level worker can perhaps contact a dozen farmers in one day on an individual basis. He can conduct two or three meetings and meet 100 to 200 people. However, he could distribute or post hundreds of leaflets, folders or other printed materials in a day. It is appropriate that mass media materials can supplement extension workers.

Basics of Mass Media

- Mass media refer to those which enable development agents or agencies to reach a large number of persons, directly (or) indirectly, with one exposure or a single source.
- Mass media channels are all those means of transmitting messages that involve a mass medium such as radio, television, and newspapers and so on, which enable a source of one or a few individuals to reach an audiences of many.
- Mass media are the means or devices of communication used to reach large number of audience at a time.
- Mass media are the tools or technologies that facilitate dissemination of information and entertainment to large numbers.

Importance of Mass Media

The media can surely act as an effective channel of two-way communication by taking problems and concerns of the people to the decision-makers. The involvement of the people with the medium and the credibility of the medium greatly increase when the medium deals with issues of immediate concern to the audience.

The role of the media has also to convince the planners, policy-makers and development economists to take up issues and concerns of development of rural areas. There is limitless scope for imparting education through a

Combination of various transmission techniques as the country abounds in traditional media of communications which can be effectively exploited to obtain optimal results.

Integration of the media: Nevertheless, there should be a strong interface between various modes of communication - the mass media, the folk norms and inter-personal exchange – each performing its specific function and supplementing and reinforcing one another. Experience suggests that the interplay of the mass media and the traditional modes of communication is effective. In India, the perspective of development through 'leap-frogging' by the technological leap was evolved by the late Vikram Sarabhai of the Indian Space Centre, Ahmedabad. Sarabhai was of the view that poor countries must find solutions "where the particular disadvantage of developing nations, which is that they have little to build on, is made an asset rather than a liability". Thus, they should "develop competence in advanced technologies and deploy them for the solution of their particular problems." Therefore, he called for highest priority being given to taking television to India's villages.

According to Sarabhai, (1975) "A national programme which could provide television to about 80 per cent of India's population during the next ten years could be of great significance to national integration, for implementing schems of economic and social development and for the stimulation and promotion of the electronics industry. It is of particular relevance to the large population living in isolated communities". It is thus that to the Nehruite concept of providing a school, a co-operative and a panchayat to every village, sarabhai added the ideas of providing a television to each village, available for community viewing, especially to the disadvantaged groups.

Social integration & communication network

Through the mass media, social integration has been sought to be achieved in our rural areasand rightly so. At the same time, our society is hierarchical and stratified and the lack of verticalmobility among the various classes, particularly the rich and the poor who are separated by varyingdegrees of social distances that have tended to widen over the years, particularly after the onset of theGreen Revolution. The various religious groups are also separated by emotional and psychologicaldistances.

In this context, the tremendous expansion in the communication infrastructure in the countryside was a welcome development. Printed text with words and figures communicate the messagethrough the sense of sight only. Radio is a mass medium, which communicates message throughsense of sound, but films and television combine both visual and auditory characteristics. Expansionof TV network has been phenomenal with the launching of INSAT IB in 1983.Impact of TV: However, very little attempt has been made to analyse and understand theprocess of communication with the rural masses. So far, the various studies conducted have beenmainly on the impact of TV on agricultural activities and farm men. They have proved that TV isinstrumental in imparting knowledge on agriculture. The audio and visual impact of TV plays a greatrole in the life of illiterate persons also. In fact, illiterates gained more than the literates. Whilecontent analysis of other TV programmes for the rural masses is still an unexplored field, very fewstudies have been made on farmwomen.India is the first country in the world to use satellite for direct telecasting to the remote villages.The Satellite Instructional Television Experiment (SITE) project was in operation for a year during975-76 and was carried out in the backward districts of Bihar, Orissa, Madhya Pradesh, Rajasthan, Andhra Pradesh and Karnataka, Later on; INSAT 2B was launched with five new satellite channelsfrom August 15, 1993. Initially, the poor had no access to the media and hence there was communitylistening/viewing.

Media and motivation

Development communication media originated primarily to carry information on agriculturehealth, population control, etc to the rural masses. However, soon it was realized that informationalone is not sufficient. Audience should be motivated to take to new practices. The media was thusgiven the additional role of motivating the audience. It was also found that unless infrastructuralsupport was available from the development agencies, new ideas could not be practiced. For example, marketing a particular variety of seed available was not possible in the given constraint. Hence, closeco-ordination between the media and development agencies was found to be very necessary and thenonly there could be success stories.Strategic role: In this connection, it is important to mention that the Kheda Communication Project was undertaken by Indian Space Research Organization (ISRO) through a memorandum ofunderstanding with Doordarshan (1975) and 600 community sets were installed in villages of Khedadistrict in Gujarat. The project received the UNESCO prize in 1985 for rural communication. Rightly, therefore, communication is viewed as a means to break out of the poverty-ignorance trap, which hasassumed the form of a vicious cycle in a developing country like ours. Luckily, communicationsystem in India has steadily expanded and the media has assumed strategic significance in the socialmarketing techniques being used for human resource development in the overall programme of ruraldevelopment.New horizons: Indeed, with communication technology so widespread resounding inenvironment, information and education is no longer the monopoly of those who can read/write orhave the means to attend institutions at specified places

During specified time. The electronic media,with its potential to reach the farthest corner of the country, can play a significant role in the field ofeducation. According to the National Policy on Education (1986), "Modern communicationtechnologies have the potential to bypass several stages and sequences in the process of developmentencountered in the earlier decades. Both the constraints of time and distance at once becamemanageable. In order to avoid structural dualism, modern educational technology must reach

Out tothe most distant areas and the most deprived sections of beneficiaries. Simultaneously with the areas of comparative affluence and ready availability."

Need for a balanced blend

With more than five lakh villages and about 5,028 blocks, the information needs of the ruralmasses are tremendous. We must make the best use of all media sources like the radio, TV, films,exhibitions, traditional media, like the folk dances, folk songs, folk dramas, etc as also talks, discussionson topics and the like. It has been found that the audio-visual medium is very effective, though itcannot replace classroom teaching altogether or take the place of a textbook. Rural newspapers havebeen utterly lacking in India. Whatever papers are available from within the rural areas; there is needfor professionalism in them. There is also need for full-fledged morning newspapers in many of therural areas. There is no gainsaying the fact that the mass media is enormously instrumental inaccelerating development in many areas of human endeavor, but its potential has been realized latein rural India.Impact of electronic media: We have yet to come to grips with the information needs of therural masses. Amongst the mass media, television certainly has made a great impact and a largeaudience has been exposed to the medium. The boost was provided by such programmes as Mahabharatand Ramayana. Improvement in infrastructural and service facilities has created awareness amongst

Villages through the messages, which are created keeping in mind the urban audience. Exposure ofrural people to the radio are the highest which is 38 per cent as compared to TV, which is 28 per cent.Radio, plays a significant role for all manner of rural people.

Media: a facilitator

There is also a great need to use extensive interviews as research tool, to make people 'speakup' on various issues. For instance, for health services to children in the rural areas, the endeavourshould be to generate awareness among the masses on the relevance of education should be togenerate awareness among the masses on the relevance of education to their daily lives and living.The media can play the role of a facilitator to create an environment congenial to literacy promotionand make people more receptive to the efforts being made by the government for educating thechildren in remote and backward rural areas.The effort should be to bring the rural communities to the state of preparedness, where theywould pressurize the government to provide the infrastructure and facilities for basic elementaryeducation as a right of each and every rural child. The media should explain that education is veryimportant and relevant in the lives of all rural people. Mass media must bring about policy changes, when and if necessary, so that there is effective use of the communication channel.

Blending technology and tradition

If modern communication technology becomes an aid for the dialogue between the classesand the masses, between the thinking sections of society and the working, it will serve as a liberatingforce, as a support to emancipation and mass creativity. But if sophisticated technology turns into asubstitute for a dialogue, into a destroyer of traditional communication among the people and betweenthe people and the governing and intellectual elite, it will turn into an engine of structural oppressionsof the masses. These words have been said by Prof. P.C.Joshi, an expert in communication technology.It would certainly be tragic if the communication revolution, which has vast potential of promotingthe realization and renewal of folk and peasant art and culture, in its actual impact promotes theirdisintegration and destruction.Software for villages: There is, therefore an utter need to develop scripts for adult and nonformalgroups of women and children in rural areas as well as urban areas, on aspects like sanitation,Mass media-concept and role of mass media in agricultural development hygiene, family planning, co-operatives, basic laws, consumer rights and so on. There is a great needfor developing effective software.In the ultimate analysis, we have to see that the mass media becomes a beneficial tool foroverall rural development. Through the media, people's lives could be enriched in various ways.And for this, variable relationship studies need to be conducted to know which factors are related andinfluence each other. The software for multimedia development and rural knowledge managementneeds to be promoted tremendously if the various ills of rural life and rural environment are to beremoved.

The message and the medium

If one were to survey the rural scene in the country, one may hear a loud uproar about thepressing need for rural marketing, which is described as fascinating, challenging, exciting and highlyprofitable, but a cursory look at the number of organizations engaged in rural marketing would provethe hollowness of such claims. Though the rural market is sufficiently large and has a lot of potentialfor the marketers interested in tapping it, it poses numerous problems. Rural markets are highlydispersed, thinly populated and scattered. They are subject to varying religious, cultural and socialpressures. This poses special problems for the marketing communicator while there is no one bestway of communicating with the rural masses. Therefore, selection of an appropriate medium ormedia f or carrying the well-designed messages to the target audience is equally important. In ruralIndia, there are age-old techniques such as word of mouth, signs, town cries, etc., having co-existedwith the numerous media developed during the last two decades or so.

No doubt, if proper types of communication channels are developed and effectively used inrural areas, they can become a very important source of disseminating knowledge and information tothe vast number of people and to far-off places in the country.

Creating awareness

The first thing it can do is to make the people aware about the very existence of the programmeschalked out for them. Most of the rural folk, particularly in the remote areas are ignorant of them. Inthe process, those who are well off manage to avail themselves of the benefits, defeating the verypurpose of the programmes. Apart from the electronic media and the press, field publicity men canmotivate the people through films and the traditional media.

Deciding priorities

The media can also send a helping hand in the identification of priority areas. The gram sabhawill draw up plans for the development of its area based on the needs of the people. The ruralreporters who make on the spot study of the village concerned can serve as a guide to the gramsabha.

Exposing lapses

Face to face talks with the beneficiaries of the programmes and representatives of PanchayatRaj institutions at various levels will bring to light the drawbacks in the implementation. Mediareports will help initiate remedial measures. While projecting the positive side, the radio and TVshould not get shy to expose the lapses.

Conclusion

Mass media refer to those, which enable development agents or agencies to reach a largenumber of persons, directly or indirectly with one exposure or a single source.The media plays a pivotal role in the national endeavour for socio-economic development.Experience suggests that the interplay of the mass media and the traditional modes of communicationare effective. The media should serve as a liberating force, as a support to emancipation and masscreativity and a beneficial tool for overall development.

FUNCTIONS OF MASS COMMUNICATION

Communication is vital for the existence of all human beings, and for the progressof humanity. It is impossible to imagine existence without communication andinteraction. Thus, communication functions as a relating tool, that createsunderstanding, facilitates work, and strengthens collective living among people.

The content of mass communication seem diverse in nature to audience as theyexpose themselves to thousands of media stimuli throughout the day and round theclock. News, reviews, commentaries, advertisements, etc can saturate the audience.Thus, the audience members select the messages according to their needs. Themajor functions of mass communication are that it informs, enriches, educates And entertains.

Functions of mass communication

The primary function of communication is to inform, instruct/ educate, entertain,and influence/persuade people to make them function smoothly and effectively.Besides, communication has a secondary function to perform as well, throughdebates and discussion, cultural promotion and integration; it fosters consensus,creativity, and understanding amongst people, groups and societies so that they livein peace and harmony.

Surveillance of environment: It is one of the most important major functions of mass communication conceptualized by Laswell. Schramm observed that in thisfunction media took on watchman's role.Surveillance is the process of monitoringthe behavior of people, objects or processes within systems for conformity toexpected or desired norms in trusted systems for security or social control.Although the word surveillance literally means "watching over" the term is oftenused for all forms of observation or monitoring, not just visual observation.

Surveillance of environment involves seeking out and then transmittinginformation about the society and all other relevant elements. Mass communicationkeeps watch on socio-political-economic events of the close and faraway localities and reaches the accounts to the audience through creative reporting. The effect ofsurveillance of environment function can be counted on individuals and society.Effects of news on individuals is:

1. Increase personal esteem: Personal esteem is built through theprestige that comes from knowing the news. There is social gain frombeing the first with the news. Quite many individuals are aware of thisgain and so collect news from various media sources.

2. Provide social base for personal interchange: Individualsremain aware in advance about what will be the subject of interactionwith their friends, how and what parts of the mass media content willbe highly desirable during conversation with others and so on.

3. Gains in and feeling for knowledge: individuals remain knowledgeable about the information on the environment and thatencourage them to seek more knowledge and satisfy their curiosity.

Information: The quality of our life would be poorer without the bit of Information we get from mass media. In the western countries, information is nowregarded as power. The more informed you are, the more powerful you become.Those who have access to information can take advantage of it in their owninterest. Mass communication provides us enormous information about theenvironment in which we live. Information such as news of war, danger, crisis, earthquake, famine, etc. is important for that helps us in taking appropriate steps tosafeguard our interests.

Entertainment: We all need entertainment to break the monotony of our hectic Stressful life and divert our attention from the troubles and tensions. Such diversionwill have a positive impact on our lives. Mass media provides a variety ofentertainment to audiences through films, TV shows, drama, dance, music, art, comedy, games, animation, etc.

Common people have the impression that mass media exist because they aredesigned to entertain. Mass media provide the audience with some sort of escapeor diversion from the realities and anxieties of daily living. Game shows, situationcomedies, serials, movies, dramas, variety shows, sport events on television screenand huge discharge of message like music, dramas (radio plays), skits, etc. fromradio-provide variety of daily entertainment material to the audiences and thelisteners. Even in print media, the use of color, fancy typefaces, the pictures andmany attractive features in the presentation style, may provide entertainment to thereaders.

Persuasion: Persuasion is an alternative term used to denote an act of influencing others. One of the most important functions of mass communication is to persuadethe other person. It is only through persuasion that one can control and governothers. But it is also possible that one may resort to persuasion with a bad motive.The receiver must be careful about the source of such persuasion. The persuasivepotential of mass communication is used heavily in both developing and developedcountries. Media are extensively used for socio-economic progress and for salespromotion of the consumer goods.

Persuasion is the process of influencing the audience members to develop afavorable attitude towards new ideas and become dissatisfied with the traditionalmethods. It seeks change in the behavior of the audience by way of substituting theold practices with the new. Similarly, when an advertiser pleads for acceptance ofhis product, he influences consumers to develop favorable attitude and behaviortowards the advertised products. On the other hand, when a political leader delivershis election speech on radio, television or in newspaper he performs a propagandajob for his party candidate. Both the electronic media and the print media havegreat persuasive potentials but that depends on the nature of the delivered message.

Persuasiveness of mass communication depends on the following factors:

1. Source Credibility- Credibility is the degree to which a communication

   Source or channel is perceived as trustworthy and competent by the receiver.The source and the medium are viewed as inseparable by the audience, socredibility of the source and the medium is often treated interchangeably.

2. Presentation of the message- the affectivity of the media communicated

   Persuasion is largely dependent on how the message is presented by thecommunicators. Clarity, brevity and preciseness are the pillars of persuasionon mass media.

3. Media factor- there is a general impression that messages having persuasive content if communicated through television, become more effective thanwhen the message is communicated through radio or print. This impressiondoes not seem to have validity in every situation with the audience for everymessage.

4. Audience factor- audience is the single most important factor to determinethe degree of persuasiveness of mass communication. The attitude of theaudiencc toward the persuasion message of the mass media largely dependsupon who gives what message in which channel; to what extent the contentof the message satisfy their needs and the intentions; to what measures the Suggestions contained in the message are in line with their pre-heldexperiences and pre-existing preferences; and, how far the message iscompatible with group norms and value system to which the audiencebelongs.

Instruction: Mass communication helps to instruct, educate and socialize theMembers of the society. Mass communication provides a fund of knowledge,expertise and skills that enable people to operate as effective members of society.It also creates awareness; give direction and opportunity to audience throughpositive impact of mass media.

Correlation: This term was used by Lasswell and is an important function of massCommunication. The role of mass media in correlation function was termed as'Forum' by Schramm. The mass communication in the role of forum is for theexchange of comments and criticism. The said function of mass communicationarouses interests through lead articles, editorials, special articles in case ofnewspapers and magazines and news commentary, current affairs in case of radioand television. The purpose is to improve the quality of usefulness of theinformation for the citizen. Forum implies a platform where exchange of viewstakes place. This function of media thus enables the audience/readers to widentheir understanding about the environment and events happening all around them.

Debate and discussion: It is through debate and discussion in media that thePublic can clarify different viewpoints on issues of public interests and arrive at ageneral agreement on matters that concern us all. Also the audience gets a chanceto present their views through debate and discussions in talk shows throughdifferent mass media.

Cultural Promotion: Media educate the people toward better living andPreserving the traditions of the society. Mass media provide an opportunity forculture to be preserved and promoted. It presents different cultures, beliefs andcustoms from different countries and helps us promote ours to all part of the world.Thus; individuals come to know one another, understand and appreciate other'sways of life and thereby develop tolerance towards one another.

Integration: Communication is a great integrating tool. Through a fund ofKnowledge or information, individuals, groups or cultures come to know oneanother, understand and appreciate other's ways of life and thereby developtolerance towards one another. It can also be the greatest disintegrating tool.

McQuail's (1983) sums up the functions of mass communication as:

Information

- Finding out about relevant events and conditions in immediate surroundings, Society and the world
- Seeking advice on practical matters or opinion and decision choices
- Satisfying curiosity and general interest
- Learning; self-education
- Gaining a sense of security through knowledge

Personal Identity

- Finding reinforcement for personal values
- Finding models of behavior
- Identifying with valued others (in the media)
- Gaining insight into oneself

Integration and Social Interaction

- Gaining insight into the circumstances of others; social empathy
- Identifying with others and gaining a sense of belonging
- Finding a basis for conversation and social interaction
- Having a substitute for real-life companionship
- Helping to carry out social roles
- Enabling one to connect with family, friends and society

Entertainment

- Escaping, or being diverted, from problems
- Relaxing
- Getting intrinsic cultural or aesthetic enjoyment
- Filling time
- Emotional rel

Importance of Mass Communication

- Mass Communication Informs
- Mass Media Entertains
    - Mass Communication Persuades
    - Mass Communication Binds
    - Popularity of Mass Media

The mass media have an important role in modern democratic society as the mainchannel of communication. The population relies on the news media as the mainsource of information and the basis on which they form their opinions and votingdecisions. Any selection of messages in the mass media will thus have a profoundeffect on the entire society.

Competition has become increasingly keen in the area of the mass media as theykeep fighting for the attention of the readers, listeners, and TV-viewers. The lifeand death of each newspaper and TV station is at stake here when the income fromadvertising and sponsoring is proportional to the number of readers or viewers. Theprinted media have problems competing with the electronic media as sources ofnews. In order to survive, they are increasingly turning to other strategies such asentertainment, titillation, scandal mongering, and spreading fear and spendingfewer resources on serious researching of news. This is not only about the survivalof the fittest of the news media; it is also about cultural selection and political selection.

The news media are the most important channels for the propagation of culture,ideas, and opinions. Most opinion formation takes place when people sit and watchnews and debates on television. Analyzing the cultural selection in the electronicinformation society, we find that an important part of the selection lies in thechoice between TV channels. Millions of lazy viewers sit in their comfortablearmchairs with remote controls in their hands zapping between action films,revivalist preachers, and commercials for a new fragrance, hardly realizing that bychoosing which cultural and political influences they expose themselves to, theyalso chose the cultural and political evolution of their country.

It is very important to analyze which selection criteria are in effect here. Theelectronic media are first and foremost pacifying. It is a relaxation machine, andthe viewer wants to be entertained. The faces on the screen are not chosen for theiropinions but for their entertainment value. TV stations do not compete onideologies but on sense impressions. An extreme example is music videos, satiatedwith fast changing sense impressions in sound as well as in pictures.

Media scientists have often discussed how much influence the media have onpeople's opinions. People tend to selectively read what they already agree with andto rationalize their preformed opinions in the face of contrary arguments.Experimental evidence seems to indicate that the mass media have little power tochange people's opinions on issues for which they already have formed a strongopinion, but they have a profound influence when it comes to setting the agendaand priming people on new issues.

We live in the age where mass communication performs certain functions that are useful to us. It is through mass communication that millions of audience is exposed to a variety of messages each day. While many consumers of media are satisfied with any single channel of mass communication, there are others who seekexposure to more that one channel. There is increasing anxiety about the adverseeffect of mass communication on society in general and individuals in particular.

In spite of limited reach, mass communication is so central to society that life seems inconceivable without mass media. They inform and shape our social life. Their influence is positive if they are able to fulfill the information and entertainment needs of the people in accordance with the existing norms, values and culture in society. Usually, the mass communication messages are positive like conveying messages for communal harmony, peace, anti-terrorism, anti-social evils, anti-drugs, etc.

The newspaper can influence the people to a large extent creating awareness and political development. Broadcasting under government control can be used for the purpose of education, social change and development of the society. Films can bring forward unhealthy social issues of our society and promote peace and harmony within communities. Thus, these mass media can contribute immensely towards nation development and social awareness.

Media like television and cable television can sometimes have bad effect in our society. The audience of these media are watching everything being telecasted in hope of entertaining themselves, be it violence, vulgarity, etc. besides this, advertising also leaves images and impact on young minds. They give children amaterialistic world, which desires unaffordable things. Also too much of television watching is creating health problems of obesity and diversion from studies, sleep and eating proper diet.

Mass communication does influence (and even reflect) social values and practices, but this influence is always in combination with a whole lot of other socio-culturaland economic and political factors. By themselves, the media have little power to influence, change and develop. For example, Hindi films may start fashions for men and women in the areas of clothes, hairstyles, manner of speech, manner of greeting, or ways of socializing. We may even go to the extreme of acting out what we see or hear in the mass media, say a violent gesture or protest, but it takes much more than film or TV to change our social and cultural values.

Competition has become increasingly keen in the area of the mass media as they keep fighting for the attention of the readers, listeners, and TV-viewers. The life and death of each newspaper and TV station is at stake here when the income from advertising and sponsoring is proportional to the number of readers or viewers. The printed media have problems competing with the electronic media as sources of news.

In order to survive, they are increasingly turning to other strategies such as entertainment, titillation, scandal mongering, and spreading fear and spending fewer resources on serious researching of news. This is not only about the survival of the fittest of the news media; it is also about cultural selection and political selection.

Tools of Mass Communication

There are different media involved in the process of mass communication. They reach every corner of the world and are very powerful. They invade even the privacy of our bedrooms. They inform, educate, entertain and persuade. They also help in the transmission of culture and perform the job of surveillance of the society. They are the mass media. The mass media are 'the vehicles of mass communication'. The prominent ones, which have become household names are newspapers, magazines, books, radio, film, television, and more recently, satellite TV and cable TV. Mass media is broadly divided into print media and electronic media. While the print media are the oldest, having a history of about five hundred years, the electronic media are products of the 20th century technological revolution.

Tools of mass communication

(a) Newspapers, Magazines, Radio, TV, Films, Records, Internet

The Newspaper*'Those who do not read the newspaper are uninformed and those who do read the newspaper are misinformed' – Mark Twain.*

It is the printed means of conveying current information. The modern Newspaper is a combination of elements from many societies and many periods of time. Even before the birth of Christ the Romans posted newssheets called 'actadiurna' in public places. The Chinese and Koreans were using wood-carved type and paper for printing several centuries before these appeared in Europe. In the sixteenth century, well after printing had come to Europe, the Venetian

Government printed a small news sheet, which could be purchased for a *gazeta* (a small coin). The use of the word 'gazette' to refer to newspapers has survived to this day. Scholars of the history of journalism suggest that many features of the modern newspaper such as the editorial, sport articles, illustrations, political columns, and even comics, were used in one place or another long before the true mass press came into being.

Brief History of Newspaper

Although printing was introduced to England in the late 1400, it was not until 1621, nearly a century and a half later that early forerunners of the newspaper began to appear. These were called 'corantos'. Their content focused on foreign intelligence, and they were not published regularly. From the beginning the publication of corantos was strongly regulated by the government. One of the interesting patterns discernible in the history of press was that the greater the extent to which a form of government is actually dependent upon favorable public opinion, the more likely it is to support a free press. When the common people play significant roles in the determination of their own political destiny, the distribution of news and political opinions is an important process.

The American and the French revolutions began germinating and the whole fabric of western society was changing. Old pattern was slowly being replaced by a new social structure within which a strong middle class would be prominent. This commercialism was dependent upon improvement in the availability of various kinds of communication media. England had many skilled writers and journalists like Addison, Steele, and Daniel Defoe. The colonial press was edited andpublished by people who were not great literary figures. They were still using the same printing technology used by Guttenberg, the first person to discover printing procedure three centuries ago. Before a true mass press could develop, a series of sweeping social change was necessary in the society. A number of printers and publishers had experimented with the idea of a cheap newspaper that could be sold to urban population.

Various approaches to this problem were tried both in England and in the United States, but without success. It remained for an obscure New York printer, Benjamin H. Day, to find a successful formula. His little paper, the *New York Sun*, began modestly enough on September 3, 1833; with the motto 'It shines for ALL'. As subsequent events proved, it did indeed shine for all. Benjamin H.Day had begun a new era in journalism within a few years would revolutionize Newspaper publishing. The Sun attracted its impressive circulation primarily by appealing to new readers who had not previously been reached by a newspaper.The newspaper had redefinition of 'News' to fit the tastes, interests, and reading skills of the less-educated level of society.

Up to that time, 'News' generally meant reports on social, commercial, or political events of genuine importance, but Benjamin Day, however, filled the paper with news of another sort-news the people in the street found exciting, catastrophe, disaster, crime, amusement, etc. the paper was vulgar, cheap, and sensational and aimed at new literate masses. By 1837 the Sun was distributing 30,000 copies daily, more than the combined total of all New York daily papers. Imitators of Dayhad started rival papers almost immediately. This penny press was successful because it had great appeal for advertisers.

More and more newspapers began to seek out the news. The role of reporter grewmore complex and specialized as papers added foreign correspondents and special news gatherers of various kinds. Reporters were sent to the scene of battles. The'surveillance' function of the press became well established. The rising demand for fresh news was met by newly formed cooperative news gathering agencies, which made use of the telegraph wires. Printing technology was making rapidstrides, moving toward ever-increasing automation. Revolving presses, with printcast in a solid metal stereotype, became capable of rolling out 10,000 and even 20,000 sheets an hour.

Papers continued to gain in popularity. In 1850 there were about two copies of a daily newspaper purchased in the United States for every ten families. This rapid growth actually continued until about the time of World War I. The last decade of the 19th century is one of special significance in the growth of the press because itwas the beginning of new kind of journalism. 'Yellow journalism' was one of the most dramatic episodes in the development of press.

Within this competitive context, brutal struggles for additional readers developed between the leaders of giant rival papers. They would fight by any means available to expand their circulation figures, which were, of course the key to increased advertising revenue and profits. Various features, devices, gimmicks, styles, and experiments were tried by each side to make its paper more appealing to the mass of readers.

Newspapers today contain many of the devices that were actually products of the rivalries of the 1890s (one of these was colour comics; an early comic character.

Called the 'Yellow Kidd' from which 'Yellow journalism' is said to derive itsname. Yellow journalism is a pejorative reference to journalism that features scandal-mongering, sensationalism, jingoism or other unethical or unprofessional practices by news media organizations or individual journalists.

Intellectuals in general were deeply wounded by yellow journalism. According to them the great new means of communication, which held forth the tantalizing potential of mass cultural and moral upliftment, was turning to be societal degeneration.

Leaders in religion, education, law and government increasingly voiced strong protests. The press lords were faced with the threat of losing public confidence. These considerations led a number of major publishers to begin to put their own houses in order. Resolution of the conflicts brought new social arrangements. Gradually, the press became less sensational and more responsible. A set of codes and norms defining its limits and responsibilities gradually became increasingly clear.

While the mass press today varies in its degree of such strict codes, the excesses of the yellow journalism is a thing of the past. Today, the newspaper is regularly published printed unbounded newsprint in broadsheet or tabloid size and serves general interests of specific communities with news, comments, features, photographs and advertisements.

Future of Newspaper

Newspaper will undoubtedly survive with some further reduction of market share. Few changes in literacy or other factors related to potential increases in readershi pare probably in the immediate future. Newspaper publication houses today have features all modern features like Internet; four colour offset printing, electronic news rooms and many such gadgets.

The future newspaper could be a tablet newspaper having liquid crystal (LCD) screen in which the contents could be collected through telephone lines or cables. Everything could be displayed on the screen at the click of a button. Another major development could be that newspaper could be customized, and people will get to read only the things, which interests them. This way the readers will exercise a lot of control on the contents of the newspaper. But only few people will get access to such electronic newspaper.

Magazine

Magazine means a 'store house' and thus has a variety of contents in it. Magazines have specific well-defined readers and thus advertiser can reach specific target people through it more effectively. The magazines are dependent on advertising and try to reach particular segments of men, women, film lovers, young generation, etc. through it. We also have trade and business magazines for people in such professions.

Newspapers, private publishing houses, societies, educational institutions or some religious organizations may publish magazines. Even some government department and political parties publish their regular magazines. Magazines are published weekly, fortnightly, monthly, half yearly and even yearly. One of the earliest magazines known was the *Journal des Scavans* founded in Paris in 1665 and initially it carried abstracts of books.

The golden age for magazines in America came in the late half of the 19th century during which channels of distribution were created as transmission network developed. Paper pulp was now cheaper; the printing processes were improved and invention of lino type facilitated automatic type setting. Yet another advancement was better photographic reproduction.

Today, we have magazines for every topic under the sun like, beauty and fashion, business and commerce, art and craft, education and career, health and grooming, photography, automobiles, electronics, science and technology, etc. magazines doplay an important role in information, education, and offer variety of subjects for entertainment of its specific target readership.

The Radio

Guglielmo Marconi of Italy invented a way to transmit sound without using wires.

By 1901, Marconi succeeded in creating a wireless communication link between Europe and North America. In 1906, Lee Forest with John Fleming perfected the'audion' or the vacuum tube, which made clear transmission of voice and musicpossible. These developments paved the way for the first ever broadcast that took place on Christmas Eve, in 1906 in USA. Later it took ten years of hard work to perfect the radio.

Radio established its place very fast in the minds of listeners. Heavy doses of infotainment including music, drama, talk shows, etc supplemented with news made radio popular overnight. Soon radio industry developed wide spreading networks and by the 1930's radio became prime mass medium. Radio broad casting was introduced in India by amateur radio clubs in Calcutta, Bombay, Madras and Lahore, though even before the clubs launched their ventures, several experimental broadcasts were conducted in Bombay.

Radio programmes may be classified into two broad groups

1. Spoken word programmes, which include news bulletins, talks, discussions, interviews, educational programmes for schools and colleges, specific audience programmes directed at women, children, rural and urban listeners, drama, radio features and documentaries.

2. Music programmes, which include disc jockey programmes, musical performances of all types and variety programmes.

Strength of the Radio

1. Radio reaches messages to illiterates, neo-literates and highly educated receivers simultaneously.

2. It is a fairly affordable to be owed by everyone.

3. The want of visual effect is compensated by sound effects, both natural and mechanical and so live effect is moderately high. The quality of voice and sound makes the communication fairly enjoyable.
4. Musical sound effect enlivens the communication and often breaks monotony.
5. It has the capacity to deliver instantaneous messages.
6. Radio does not require captivity. Listeners can receive messages even when they are working. Farmer may listen to farm programmes while working in fields, a busy executive may listen to news bulletin even while driving or ahousewife may listen to her favorite programme even while working in the kitchen.
7. Radio does not require power line for operation and so people in remote villages devoid of power lines can also receive messages from this medium.
8. Once a transistor radio is purchased, messages flow constantly and no cost is involved for reception of messages.

Weakness of the Radio

1. Communication through radio lacks visual component and so does not demonstrate but suggests
2. Spoken messages are subject to interpretation of listeners according to their imagination, experience and predisposition; hence possibility of misinterpretation is very high.
3. Listeners need lot of imagination and therefore understanding of message depends largely on the characteristics of the receivers.
4. Communication is time limited and presents tiny fragments of topics in ahaphazard mosaic. The medium has also limitations because of its audio nature.
5. Receivers cannot put off listening parts of message for subsequent listening at their convenience.

Future of Radio

Radio's future is a mystery. It is not easy to predict the future of radio. The future of radio would depend on changing regulatory scenarios, technological developments and change of listener's appeal.

Radio's current localization and specialized programming will continue. Technologically, radio transmission will improve greatly. FM will continue to grow faster and bigger.

Radio listeners have indeed grown manifold, and the network is expanding a great deal an now it offers a daily service for many hours transmitting news, comments, songs, music, comedies, thrillers, sports, besides special programmes for children,youth and farmers. One of the best advantages that radio has over other media is that it can serve and entertain an audience, which is otherwise occupied. For example, people can listen to it while working at home, in the fields and factories and even while travelling.

Television

Unlike other forms of mass media, television has become one of the most powerful media of Mass communication. With a modest beginning in the1930s, it has grown into a massive network of mass information and mass entertainment in today's world. The attraction of the 'visual ness' of the medium makes people remain glued to the TV set for hours. Television captures our imagination and is the most complete and dramatic of all mass media. In addition to providing news and events, television also packages fiction, drama, culture, economy and many other things. Thus, this idiot box (because it provides everything on a platter and we need not do any thinking) has been increasing its hold on us.

History of Television

The inventions and discoveries in the late 1990s and early twentieth century, which gave us radio, films and the telephone, also lead to the invention of the television.

Vladimir Zworykin, an American scientist, who developed an all-electronic Television system in 1923 and perfected it by 1928, took the first big step in the development of TV. However, only experimental TV broadcasts were conducted in the early days. In 1938, TV sets became widely available and since then there is no looking back. In India, television arrived with small scale experimental telecasting from Delhi in 1959. Slowly the half hour programme experiment grew. While Doordarshan was the only channel available through 1980, the TV in India has completely changed with the arrival of private TV channels.

Strength and weakness of Television

It has all the strength of radio except that it needs captive audience, has not attained portability and miniaturization and needs power line for inexpensive working. TV viewing is essentially a family affair and so helps family unity. It isfar from slow and availability is constant. Repetition of message does not in curexpense except nominal fees for cable connections. Television like radio, is in all sense a 'now' medium.

Television gives cursory overview of the events and is never capable of providing in-depth analysis and reporting. It is severely time limited and presenting fragments of topics. Television programmes skip and jumps demands constant change of mental gears from programmes to commercial, from documentary to cartoon and news. Also contents are high in entertainment and low in information. Power line is needed for its inexpensive operation and battery operated television sets are most expensive.

Films

Films refer to all documentary, educational, feature, informational and advertisement cinemas. A film is considered a mass medium because it reaches to a very large audience. It is not as immediate as newspaper, television or radio, but reach a large number of people over a long period of time. Film does not have a well-defined audience like other major mass media. It also lacks the over whelming presence as enjoyed by radio, TV and newspapers, etc. but still it has one very big advantage that is it commands attention. Once inside the darkened theatres,audience members forget the outside world and become captive to the charm of the film. The larger than life format of film (from 16mm, 35mm, 70mm, cinema scope to the most recent I-max format) provides the film producer absolute control over *Emphasis, order of presentation, continuity, dramatic effect and timing.* Thus film has become a superb medium of entertainment, instruction and persuasion. When individual photographs or visuals are shown one after the other at a very fast rate, then we get an illusion of motion or movement. Cinema works on the principles of 'persistence of vision', which means that the eye retains an image for fleeting seconds after it is gone. Efforts had started very early to create illusion of motion. Many devices were invented for this purpose. Long back Leonardo daVinci developed the camera obscura. In 1671, Kircher developed the 'magic lanterns'.

Photography and projection were united when Stanford developed the'Zoo praxinoscope'. Soon Thomas Alva Edison invented the electric bulb and many other such developments took place. Then came the Lumiere brother's who produced and started having commercial shows of short shoot and show films.

Soon others followed and by the beginning of the 20th century, film became the second mass medium after newspapers. Very soon films became a form of family entertainment. Movie theaters opened every where and people wanted to see more interesting contents. So feature films came into existence. The combination of allthese factors made cinema a booming industry.

Film has had an enormous impact on the audiences. One reason is it is not imposed. It does not come to us and instead we go to theatres to watch films.Usually the films deal with universal themes so language barrier is minimal and we can thoroughly enjoy film of another language if we like the theme.

India is in fact the largest producer of feature films in the world. Commercial cinema is all glamour and fantasy. The usual ingredients are sex, songs, dances, crime, fights, melodrama, and comedy, all bordering on unreality. Then also, thesefilms set trends in styles and tastes, dominate the popular radio and television entertainment programmes, provide spicy reading material for film magazines, which are published in large numbers.

Strength and weakness

Cinemas are replica of dramas in natural settings and so influence audience. Evenmyths are depicted as if they are real. Cinema is an audio-visual medium and is rich in live effect and demonstrates as well as suggests. Details are extensive through sound, music, visual effects, and skillful production, editing and role-playing. Dramatization of the presentation sets tempo and mood of the audience. And most significant attribute of film is that it reaches messages to illiterates, neoliterates, and moderately educated and highly educated people having basic visual literacy.

Selection of sets and props in films sometimes confuse the audience, makes the communication abstract and creates misunderstanding. Crime and obscene adversely affect the society and so realistic censorship is required for the welfare of the society. Cost of exposure is moderately high especially in case of commercial cinema.

Future of films

In its century old existence cinema has faced few shakes ups. First it was television. Skeptics thought no one would watch films in theatres, as so much was available on television with in the comforts of one's home. But soon it was found that TV depends too heavily on films and films form a considerable part of TV programming.

Then came videocassettes, now it is videodiscs. These eliminate the necessity of "going out" to the theatres to be entertained. But the fact that cinema is thriving proves that mass media share a symbiotic relationship and are not mutually destructive. Of course, cinema is not sitting idle. It is facing the threat posed by other media head-on.

Highly decorated theatre halls complete with shopping complexes, are now attracting more audience. Multiplexes, like PVR Delhi are another way of filmfighting back. Then there are 70 mm and cinema scope. Faster frame rate is anothernovelty. Hollywood has started delivering films to theatre halls over satellite.

Another recent innovation is the *I- max* screens, which are ten times larger than the traditional 35 mm screen. Dolby stereo system, 16-track recording, etc. also have added more allure to films. Interactive films, where audience member can have a say about how a film should end, is another novel way of attracting more audience.

So it can be safely concluded that film, as a medium of entertainment and communication and as an industry, would continue to grow and hold an importantpart in our social system.

Records

A gramophone record (or simply record) is an analogue sound recording medium consisting of a flat disc with an inscribed modulated spiral groove starting near the periphery and ending near the center of the disc. Gramophone records were the primary medium used for commercial music reproduction for most of the 20th century. They replaced the phonograph cylinder in the 1900s, and although they were supplanted in popularity in the late 1980s by digital media, they continue to be manufactured and sold as of 2007. Gramophone records remain the medium of choice for many audiophiles and music a ficionados, especially in the electronic agenre.

In 1878, Thomas Edison independently built the first working phonograph, a tin foil cylinder machine, intending to use it as a voice recording medium, typically for office dictation. The phonograph cylinder dominated the recorded sound market beginning in the 1880s.

Materials

Early disc records were originally made of various materials including hard rubber.From 1897 onwards, earlier materials were largely replaced by a rather brittle formula of 25% "shellac" (a material obtained from the excretion of a southeast Asian beetle), a filler of a cotton compound similar to manila paper, powdered slate, and a small amount of a wax lubricant. The mass production of shellac records began in 1898 in Germany. Shellac records were the most common until the 1950s. Unbreakable records, usually of celluloid (an early form of plastic) on apasteboard base, were made from 1904 onwards, but they suffered from a nexceptionally high level of surface noise.

In the 1890s the early recording formats of discs were usually seven inches From1903 onwards, 12-inch records (30.5cm) were also commercially sold, mostly ofclassical music or operatic selections, with four to five minutes of music per side.

Such records were usually sold separately, in plain paper or card board sleeves that may have been printed to show the producer or the retailer's name and, starting in the 1930's, in collections held in paper sleeves in a cardboard or leather book, similar to a photograph album, and called *record albums*. Empty record albumswere also sold that customers could use to store their records in. The Long-P laying

Records (LPs) usually come in a paper sleeve within a colour printed card jacket which also provides a track listing?

Records

A gramophone record (or simplyrecord) is an analogue sound recording medium consisting of a flat disc with an inscribed modulated spiral groove starting near the periphery and ending near the center of the disc. Gramophone records were the primary medium used for commercial music reproduction for most of the 20th century. They replaced the phonograph cylinder in the 1900s, and although they were supplanted in popularity in the late 1980s by digital media, they continue to be manufactured and sold as of 2007. Gramophone records remain the medium of choice for many audiophiles and music aficionados, especially in the electronica genre.

Time. The global music industry has been estimated worth over $40 billion, out of Which India's share is $250 million?

'Remix albums' had first turned into a rage in the 80s and 90s because of the growing interest among young music listeners in the cities in disco rhythms and excellent sound quality. Also, their interest in old film songs has remained intact.This generally involves first re-recording the rhythm track and then superimposinga vocal track on it. Almost, all music majors in India like T-Series, Tips andVenus, are involved in the business of're-mixes'.

Internet

There is a rising use of personal computers by individuals at home and in offices. Today, average person has the skill to use basic computer in daily life. As the essence of mass communication is that professional communicators operate the media to a large and heterogeneous audience, a computer network in which peoples end messages to each other will be a different kind of process altogether. However, where records of memos, messages, and transactions are done, this medium would be considered to be a medium of mass communication. A more likely prospect is that new mass media will develop by coupling computers to modern variants of cable television. In fact, experimental media using this technology have already come into use.

The Internet is the worldwide, publicly accessible network of interconnected computer networks that transmit data by packet switching using the standard Internet Protocol (IP). It is a "network of networks" that consists of millions of smaller domestic, academic, business, and government networks, which together carry various information and services, such as electronic mail, online chat, file transfer, and the inter linked Web pages and other documents of the World WideWeb.

The prevalent language for communication on the Internet is English. This may bea result of the Internet's origins, as well as English's role as the lingua franca. It may also be related to the poor capability of early computers to handle characters other than those in the basic Latin alphabet.

The Internet is allowing greater flexibility in working hours and location,especially with the spread of unmetered high-speed connections and Web applications.

The Internet can now be accessed virtually anywhere by numerous means. Mobile phones, datacards, handheld game consoles and cellular routers allow users to connect to the Internet from any where there is a cellular network supporting that device's technology.

Common uses of the Internet

E-mail-The concept of sending electronic text messages between parties in a way analogous to mailing letters or memos predates the creation of the Internet. Even today it can be important to distinguish between Internet and internal e-mail systems. Internet e-mail may travel and be stored unencrypted on many other machines and networks out of both the sender's and the recipient's control.

The World Wide Web-Through keyword-driven Internet research using search engines, like Google, millions worldwide have easy, instant access to a vast and diverse amount of online information. Compared to encyclopedias and traditional libraries, the World Wide Web has enabled a sudden and extreme decentralization of information and data.

Remote access-The Internet allows computer users to connect to other computers and information stores easily, wherever they may be across the world. This is encouraging new ways of working from home, collaboration and information sharing in many industries. An accountant sitting at home can audit the books of acompany based in another country.

Collaboration-The low-cost and nearly instantaneous sharing of ideas, knowledge, and skills has made collaborative work dramatically easier. Not only can a group cheaply communicate and test, but the wide reach of the Internet allows such groups to easily form in the first place, even among niche interests. Internet 'chat', whether in the form of IRC 'chat rooms' or channels, or via instant messaging systems allow colleagues to stay in touch in a very convenient way when working at their computers during the day. Messages can be sent and viewed even more quickly and conveniently than via e-mail.

Media has been persuaded to come out of their shell to give a personal touch to the otherwise impersonal mass media programmes.

Behavioral changes are most easily brought about by personal interaction and traditional folk media are personal forms of communication, of entertainment. These forms of art are a part of the way of life of a community and provide acceptable means of bringing development issues into the community on its ownterms. Traditional folk media are playing a meaningful role in the affairs ofdeveloping countries like Asia and Africa.

Traditional folk media are rich in variety, readily available and economically viable. Men and women of different age groups relish them. They command the confidence of the rural masses, as they are live. They are theme-carriers by nature,not simply as vehicles of communication but as games of recreating and sharing a common world of emotions, ideals and dreams. Also, traditional folk media are in a 'face to face situation' between the communicator and the receiver of the message, a situation which energies discussion that may lead to conviction and motivation.

In developing country like India, traditional folk media have proven to be successful mass-motivators. During the years before the advent of the sound and sight channels of mass media, the traditional folk media not only reflected the joys and sorrows of the people, but also inspired the mass during the times of stress and strain. They played a significant role in the freedom movement of India. Since the country has achieved independence, selected folk media have been effectively harnessed for communication of new 'development messages'. Mass media have extended the area of coverage of a folk performance, while traditional folk media,with their inspiring color and costume, dance and music; have enriched the content of the mass media channels.

Traditional folk performances like ritualistic dances, religious songs and mythology based rural plays, though highly popular, have proved unsuitable to absorb and reflect new messages on population, health and hygiene. Communicators, therefore, have to test different categories of folk performances to identify the ones that are flexible enough to absorb development messages to meet the contemporary needs. Flexibility is the most important factor, which determines the viability of a folk medium for rural communication.

Some traditional folk media in India are: Tamasha, Nautanki, Jatra, Bhavai, Puppetry, Ramlila & Raaslila, Street Theatre, Pawala, Keertana, and others.

There are different media involved in the process of mass communication. They reach every corner of the world and are very powerful. They invade even the privacy of our bedrooms. They inform, educate, entertain and persuade. They also help in the transmission of culture and perform the job of surveillance of the society. They are the mass media. The mass media are 'the vehicles of mass communication'. For the common men advertising means television commercials, radio jingles and print advertisements. Advertising can be defined as a paid dissemination of information through a variety of mass communication media to motivate a desired action. According the 'concise oxford dictionary' the verb, 'to advertise' means: to make generally or publicly known.

Public relations(PR) is the business, organizational, philanthropic, or social function of managing communication between an organization and its audiences.There are many goals to be achieved by the practice of public relations, including education, correcting a mistruth, or building or improving an image.

Public Relation is a management function that involves monitoring and evaluating public attitudes and maintaining mutual relations and understanding between an organization and its public. Public could include shareholders, government,consumers, employees and the media. It is the act of getting along with people we constantly come in touch with. PROs ensure internal cohesion in the company by maintaining a clear communications network between the management and employees.

Public affairs programming, a broadcasting industry term, refers to programming which focuses on matters of politics and public policy. Among commercial broadcasters, such programs are often only to satisfy regulatory expectations, and hence are scheduled at times when few listeners or viewers are tuned in (or even awake).

Folk and traditional media continue to play an important role in our society and the new electronic media are used to popularize some of the folk arts. Ingredients of folk media are given special projection in the mass media and as such folk media are being used in development communication (to bring about attitudinal and behavioral changes of the people) and advertising. Messages on issues like agricultural development, primary health care and nutrition; education, women and child rights are projected through the folk media. The great majority in the rural areas enjoys, performances of the folk artists as a relief from the myriad of life. Many of them simply do not have access to modern forms of entertainment.

Uttar Pradesh had the largest number of daily newspapers (285), followed by Maharashtra (197). Daily newspapers are published from all the states. However, no circulation details were made available from the union territories of Andaman & Nicobar Islands, Dadra & Nagar Haveli, and Lakshadweep. Dailies from UttarPradesh with a total circulation of 1,11,45,976 copies were at the top followed by Maharashtra with 83,46,863 and Delhi 80,79,842 copies respectively.

A notable feature was that Orissa achieved the distinction of publishing newspapers in 17 major languages (bilingual and multilingual publications have not been included in the 17 major languages).

Hindustan Times, published from Delhi and printed at New Delhi, Gurgaon, Noida, Chandigarh, Bhopal and Jaipur, was the largest circulated single editiondaily with 11,75,339 copies followed by Ananda Bazar Patrika, published from Kolkata with 11,30,167 copies. The Times of India, published from New Delhi and printed at New Delhi and Sahibabad came third with 11,02,772 copies.

The Times of India having eight editions with a combined circulation of 27,71,669 copies came first among multi-edition dailies. Dainik Jagran in Hindi having 15 editions, claiming a combined circulation of 24,83,432 copies stood second. Dainik Bhaskar (Hindi) with 16 editions occupied the third position.

India has four news agencies namely, the Press Trust of India (PTI), United News of India (UNI), Samachar Bharti and Hindustan Samachar. Newspapers and magazines in India are independent and largely privately owned. About 5,000 newspapers, 150 of them major publications, are published daily in nearly 100 languages. Over 40,000 periodicals are also published in India. The periodicals specialize in various subjects but the majority of them deal with Subjects of general Interest.

Some major Newspapers in India

The Times of India

The Times of India is one of the leading newspapers of India and is very old. Some of the features of the paper are news from various Indian cities, sports, business (Indian and international), entertainment, health, and the world. As a policy it lays more stress on human-interest stories. It also carries features like Sports, Entertainment, Stocks, Infotech, Editorials, Interviews, Letters, Cross words, and Horoscopes etc. The Times group also owns The Economic Times, Femina, Filmfare, etc.

The Economic Times

The Economic Times is one of the India's leading business newspapers. It carries news about the Economy, Companies, Infrastructure, Trends in the Economy, Finance, Stocks, Forex and Commodities, news from around the world and from theworld of politics besides Editorial and Features. The magazine section has Investor's Guide, Brand Equity and Corporate Dossier. It is part of the Times Group.

Indian Express

The Indian Express is another leading newspaper of the country. This group's publications includes the Indian Express, the Financial Express, Loksatta, and Screen and Express Computer.

India Today

India Today is part of Living Media India Ltd, one of India's leading names in news and publishing which includes the television production group of Aaj Tak and Headlines Today, along with Business Today, Computers Today, India Today Hindi, India Today Plus, Teens Today and Music Today. It contains everything from hardcore political and business news and analysis to the movies, music, and art, books, computers, and lifestyle and business resources.

❑❑❑

Chapter 3

Theories of Mass Communication

The study of communication and mass media has led to the formulation of many theories: structural and functional theories believe that social structures are real and function in ways that can be observed objectively; cognitive and behavioral theories tend to focus on psychology of individuals; interactionist theories view social life as a process of interaction; interpretive theories uncover the ways people actually understand their own experience; andcritical theories are concerned with the conflict of interests in society and the way communication perpetuates domination of one group over another .

The earliest theories were those propounded by Western theorists Siebert, Paterson and Schramm in their book *Four Theories Of the Press* (1956). These were termed "normative theories" by McQuail in the sense that they "mainly express ideas of how the media ought to or can be expected to operate under a prevailing set of conditions and values." Each of the four original or classical theories is based on a particular political theory or economic scenario.

Classical Theories

Authoritarian Theory

According to this theory, mass media, though not under the direct control of the State, had to follow its bidding. Under an Authoritarian approach in Western Europe, freedom of thought was jealously guarded by a few people (ruling classes), who were concerned with the emergence of a new middle class and were worried about the effects of printed matter on their thought process. Steps were taken to control the freedom of expression. The result was advocacy of complete dictatorship. The theory promoted zealous obedience to a hierarchical superior and reliance on threat and punishment to those who did not follow the censorship rules or did not respect authority. Censorship of the press was justified on the ground that the State always took precedence over the individual's right to freedom of expression.

This theory stemmed from the authoritarian philosophy of Plato (407 - 327 B.C), who thought that the State was safe only in the hands of a few wise men. Thomas Hobbes (1588 - 1679), a British academician, argued that the power to maintain order was sovereign and individual objections were to be ignored. Engel, a German thinker further reinforced the theory by stating that freedom came into its supreme right only under Authoritarianism.

The world has been witness to authoritarian means of control over media by both dictatorial and democratic governments.

Libertarianism or Free Press Theory

This movement is based on the right of an individual, and advocates absence of restraint. The basis of this theory dates back to 17th century England when the printing press made it possible to print several copies of a book or pamphlet at cheap rates. The State was thought of as a major source of interference on the rights of an individual and his property. Libertarians regarded taxation as institutional theft. Popular will (vox populi) was granted precedence over the power of State.

Advocates of this theory were Lao Tzu, an early 16th century philosopher, John Locke of Great Britain in the17th century, John Milton, the epic poet ("*Aeropagitica*") and John Stuart Mill, an essayist ("*On Liberty*"). Milton in*Aeropagitica* in 1644, referred to a self righting process if free expression is permitted "let truth and falsehood grapple." In 1789, the French, in their Declaration of the Rights of Man, wrote

"Every citizen may speak, write and publish freely." Out of such doctrines came the idea of a "free marketplace of ideas." George Orwell defined libertarianism as "allowing people to say things you do not want to hear". Libertarians argued that the press should be seen as the *Fourth Estate* reflecting public opinion.

What the theory offers, in sum, is power without social responsibility?

Social Responsibility Theory

Virulent critics of the Free Press Theory were Wilbur Schramm, Siebert and Theodore Paterson. In their book *Four Theories Of Press,* they stated "pure libertarianism is antiquated, outdated and obsolete." They advocated the need for its replacement by the Social Responsibility theory. This theory can be said to have been initiated in the United States by the Commission of The Freedom Of Press, 1949. The commission found that the free market approach to press freedom had only increased the power of a single class and has not served the interests of the less well-off classes. The emergence of radio, TV and film suggested the need for some means of accountability. Thus the theory advocated some obligation on the part of the media to society. A judicial mix of self regulation and state regulation and high professional standards were imperative.

Social Responsibility theory thus became the modern variation in which the duty to one's conscience was the primary basis of the right of free expression.

Soviet Media/Communist Theory

This theory is derived from the ideologies of Marx and Engel that "the ideas of the ruling classes are the ruling ideas". It was thought that the entire mass media was saturated with bourgeois ideology. Lenin thought of private ownership as being incompatible with freedom of press and that modern technological means of information must be controlled for enjoying effective freedom of press.

The theory advocated that the sole purpose of mass media was to educate the great masses of workers and not to give out information. The public was encouraged to give feedback as it was the only way the media would be able to cater to its interests.

Two more theories were later added as the "four theories of the press" were not fully applicable to the non-aligned countries of Asia,

Africa and Latin America, who were committed to social and economic development on their own terms. The two theories were:

The underlying fact behind the genesis of this theory was that there can be no development without communication. Under the four classical theories, capitalism was legitimized, but under the Development communication theory, or Development Support Communication as it is otherwise called, the media undertook the role of carrying out positive developmental programmes, accepting restrictions and instructions from the State. The media subordinated themselves to political, economic, social and cultural needs. Hence the stress on "development communication" and "development journalism". There was tacit support from the UNESCO for this theory. The weakness of this theory is that "development" is often equated with government propaganda.

Democratization/Democratic Participant Media Theory

This theory vehemently opposes the commercialization of modern media and its top-down non-participant character. The need for access and right to communicate is stressed. Bureaucratic control of media is decried.

Magic Bullet/Hypodermic Needle/Stimulus Response Theory

Before the first World War, there was no separate field of study on Communication, but knowledge about mass communication was accumulating. An outcome of World War I propaganda efforts, the Magic Bullet or Hypodermic Needle Theory came into existence. It propounded the view that the mass media had a powerful influence on the mass audience and could deliberately alter or control peoples' behaviour.

Klapper (1960) formulated several generalizations on the effects of mass media. His research findings are as follows: "*Mass-media ordinarily does not serve as a necessary and sufficient cause of audience effect, but rather functions through a nexus of mediating factors and influences. These mediating factors render mass-communication as a contributory agent in a process of reinforcing the existing conditions.*"

The main mediating factors which he considers responsible for the functions and effects of mass communications are - selective exposure i.e., people's tendency to expose themselves to those mass communications which are in agreement with their attitudes and interests; and selective perception and retention i.e., people's inclination to organize the meaning of mass communication messages into accord with their already existing views.

Two Step Flow Theory

In the early 40"s, before the invention of television, Lazarsfeld, Berelson and Goudet conducted an American survey on mass campaigns. The study revealed that informal social relationships had played a part in modifying the manner in which individuals selected content from the media campaign. The study also indicated that ideas often flowed from the radio and newspapers to opinion leaders and from them to the less active sections of society.Thus, informal social groups have some degree of influence on people and mould the way they select media content and act on it.

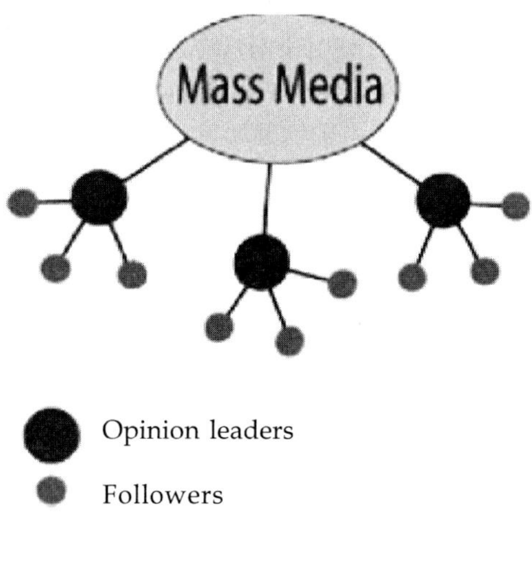

Figure 3.1
Source: CIA Advertising at www.ciadvertising.org/ortega/Theories.htm (used by permission)

One Step Flow Theory

This theory simply stated that mass communication media channels communicate directly to the mass audience without the message being filtered by opinion leaders.

Multi Step Flow Theory

This was based on the idea that there are a number of relays in the communication flow from a source to a large audience.

Uses andf Gratification Theory

This theory propounded by Katz in 1970, is concerned with how people use media for gratification of their needs. An outcome of Abraham Maslow's *Heirarchy of Needs,* it propounds the fact that people choose what they want to see or read and the different media compete to satisfy each individual"s needs.

In the hierarchy of needs, there are five levels in the form of a pyramid with the basic needs such as food and clothing at the base and the higher order needs climbing up the pyramid. The fulfillment of each lower level need leads to the individual looking to satisfy the next level of need and so on till he reaches the superior-most need of self-actualization.

Source : William G Huitt - Valdosta University at chiron. valdosta. edu/whuitt/ (used by permission)

The Uses and Gratifications approach reminds us that people use media for many purposes. As media users become increasingly confronted with choices, this approach should direct our attention to the audience. Lull's television research found that families used television for communication facilitation, relationship building, intimacy, and for structuring the day. In general researchers have found four kinds of gratifications:

1. Information - we want to find out about society and the world- we want to satisfy our curiosity. This would fit the news and documentaries which both give us a sense that we are learning about the world.
2. Personal Identity - we may watch the television in order to look for models for our behaviour. So, for example, we may identify with characters that we see in a soap. The characters help us to decide what feel about ourselves and if we agree with their actions and they succeed we feel better about ourselves.
3. Integration and Social Interaction - we use the media in order to find out more about the circumstances of other people. Watching a show helps us to empathize and sympathize with the lives of others so that we may even end up thinking of the characters in programme as friends.
4. Entertainment - sometimes we simply use the media for enjoyment, relaxation or just to fill time.

Riley and Riley (1951) found that children in peer groups used adventure stories from the media for group games while individual children used media stories for fantasizing and daydreaming. The study thus found that different people use the same messages from the media for different purposes.

Katz replaced the question "what do media do to people?" with the question "what do people do with the media?" Katz, Gurevitch & Hass found that the media are used byindividuals to meet the following specific needs:

Cognitive needs (acquiring information, knowledge and understanding);

Affective needs (emotional, pleasurable experience);

Personal integrative needs (strengthening self image);

Social integrative needs (strengthening self image);

Tension release needs (escape and diversion)

McQuail, Blumler and Brown suggested the following individual needs categories:

1. Diversion (emotional release)
2. Personal Relationships (substitute of media for companionship).
3. Personal identity or individual psychology (value reinforcement, self understanding.)
4. Surveillance (information that may help an individual accomplish tasks.)

B. Rubin and Bantz (1989) studied the uses and gratifications of "new technology" by examining VCR use. They found the following motives for VCR use:

1. Library storage of movies and shows
2. Watching music videos
3. Using exercise tapes
4. Renting movies
5. Letting children view
6. Time-shifting
7. Socializing by viewing with others
8. Critical viewing including TV watching and studying tapes

Spiral of Silence Theory

Propounded by Elisabeth Noelle-Neumann, this theory states that the media publicizes opinions that are mainstream and people adjust their opinions according to their perceptions to avoid being isolated. Individuals who perceive their own opinion as being accepted will express it, whilst those who think themselves as being a minority, suppress their views. Innovators and change agents are unafraid to voice different opinions, as they do not fear isolation.

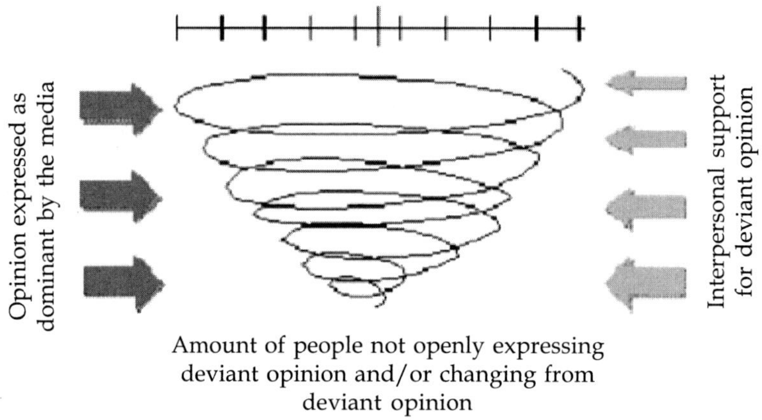

Noelle-Neumann's Spiral of Silence

Figure 3.2

Source: CCMS-Infobase at http://www.cultsock.ndirect.co.uk/

Consistency Theories (1950s)

Festinger formulated the consistency theories that talked about people"s need for consistency in their beliefs and judgements. In order to reduce dissonance created by inconsistencies in belief, judgments and action people expose themselves to information that is consistent with their ideas and actions, and they shut out other communications.

McCombs and Shaw's Agenda Setting Theory

This theory puts forth the ability of the media to influence the significance of events in the public's mind. The media set the agenda for the audience's discussion and mentally order and organize their world. The theory is consistent with a "use and gratification" approach. McCombs and Shaw assert that the agenda-setting function of the media causes the correlation between the media and public ordering of priorities. The people most affected by the media agenda are those who have a high need for orientation.

Media Dependency Theory

Developed by Ball-Rokeach and DeFluer, the key idea behind this theory is that audiences depend on media information to meet needs and reach goals, and social institutions and media systems interact with audiences to create needs, interests, and motives in the person. The degree of dependence is influenced by the number and centrality of information functions and social stability. Some questions that this theory raised were:

Do media create needs?

Do people turn to media to achieve gratification and satisfy needs?

Are media needs personal, social, cultural, political, or all of these?

"The media are our friends"??

Stephenson"s Play Theory

Play is an activity pursued for pleasure. The daily withdrawal of people into the mass media in their after hours is a matter of subjectivity. The effect of mass communication is neither escapism nor seducing the masses. Rather it is seen as anti-anxiety producing, and are regarded as communication-pleasure.

Modeling Behaviour Theory

Behaviors which are modeled from media experiences can become habitual if found useful and/or if they are reinforced in the environment. This is not about violent or criminal behavior.

Stalagmite Theories

These theories suggest that mediated experiences induce long term effects that are very difficult to measure. The effects are like stalagmite drippings building up over time. Meaning Theory and the Cultivation Theory are two of the most significant Stalagmite theories.

Meaning Theory

Media experiences mould meanings by putting things in a particular framework. Does "NYPD Blue" depict the real world of New York City police detectives? Questions like this are coming from a Meaning Theory focus on media.

Cultivation Theory

George Gerbner tried to determine the influence of television on viewers" ideas of the environment they lived in. He found that dominance of TV created a common view of the world and that it homogenized different cultures. TV portrayed the society as a bad place to live in leading to people becoming distrustful of the world. Over time, particular symbols, images, messages, meanings become dominant and are absorbed as the truth. Cultural stereotypes, ways of assessing value and hierarchies are established.

Diffusion of Innovations Theory

Pioneered in 1943 by Bryce Ryan and Neil Gross of Iowa State University this theory traces the process by which a new idea or practice is communicated through certain channels over time among members of a social system.The model describes the factors that influence people's thoughts and actions and the process of adopting a new technology or idea.

Social Learning Theory

Formulated by Albert Bandura at Stanford University, this specifies that mass-media messages give audience members an opportunity to identify with attractive characters that demonstrate behavior, engage emotions, and allow mental rehearsal and modeling of new behavior. The behavior of models in the mass media also offers vicarious reinforcement to motivate audience members' adoption of the behavior.

Baran and Davis (2000) classify mass communication theories into three broad categories:

1. Microscopic theories that focus on the everyday life of people who process information - for example, uses and gratifications, active audience theory, and reception studies;
2. Middle range theories that support the limited effects perspective of the media - for example, information flow theory, diffusion theory, and
3. Macroscopic theories that are concerned with media's impact on culture and society - for example, cultural studies theory.

Theories of mass communication have always focused on the "cause and effects" notion, i.e. the effects of the media and the process leading to those effects, on the audience's mind. Harold Lasswell and

Berelson have succinctly expressed this idea. Lasswell's essential question is timeless (1949): *"Who says what in what channel to whom with what effects?"* Berelson said: *"Some kinds of communication, on some kinds of issues, brought to the attention of some kinds of people, under some kinds of conditions, have some kinds of effects."* (1949).

Wilbur Schramm stated: *"In fact, it is misleading to think of the communication process as starting somewhere and ending somewhere. It is really endless. We are little switchboard centers handling and rerouting the great endless current of information.... "* (Schramm W.1954) quoted in McQuail & Windahl (1981).

The Osgood and Schramm circular model emphasizes the circular nature of communication.

The participants swap between the roles of source/encoder and receiver/decoder.

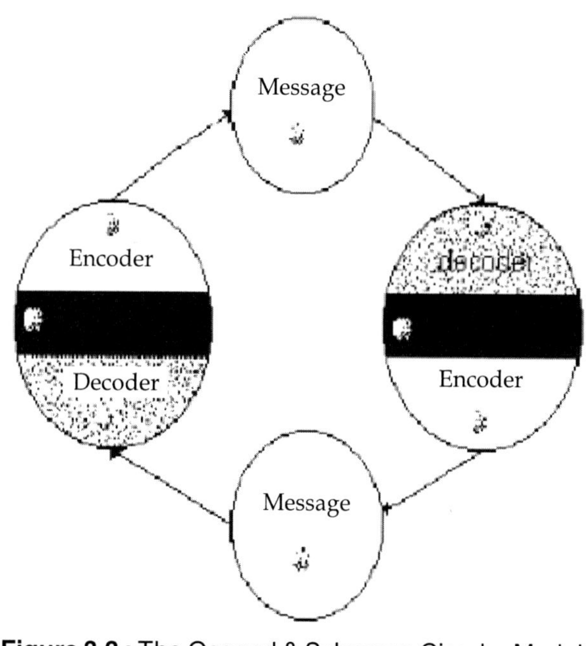

Figure 3.3 : The Osgood & Schramm Circular Model
Source : CCMSInfobaseathttp://www.cultsock.ndirect.co.uk/

Gerbner's General Model

Gerbner's General Model also emphasizes the dynamic nature of human communication.

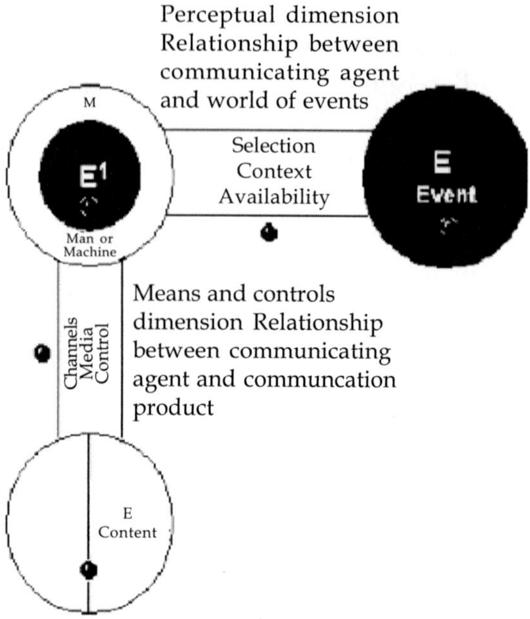

Gerbner's General Model
Please click a button for further information

Figure 3.4
Source: CCMS-Infobase at http://www.cultsock.ndirect.co.uk/

The Shannon-Weaver Model

Shannon and Weaver produced a general model of communication known after them as the Shannon-Weaver Model. It involved breaking down an information system into sub-systems so as to evaluate the efficiency of various communication channels and codes. They propose that all communication must include six elements:

Source

Encoder

Channel

Message

Decoder

Receiver

This model is often referred to as an "information model" of communication. A drawback is that the model looks at communication as a one-way process. That is Remedied by the addition of the feedback loop. Noise indicates those factors that disturb or otherwise influence messages as they are being transmitted.

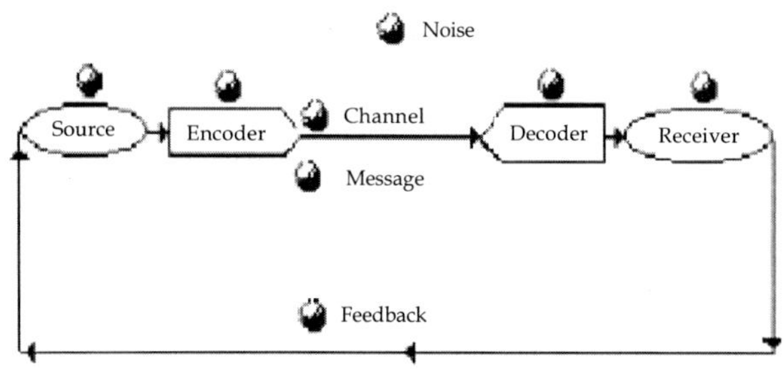

Figure 3.5 : The Shannon-Weaver Model

Berlo's S-M-C-R Model

Berlo"s SMCR (Source, Message, Channel, and Receiver) model focuses on the individual characteristics of communication and stresses the role of the relationship between the source and the receiver as an important variable in the communication process. The more highly developed the communication skills of the source and the receiver, the more effectively the message will be encoded and decoded.

Berlo's model represents a communication process that occurs as a SOURCE drafts messages based on one's communication skills, attitudes, knowledge, and social and cultural system. These Messages are transmitted along Channels, which can include sight, hearing, touch, smell, and taste. A Receiver interprets messages based on the individual's communication skills, attitudes, knowledge, and social and cultural system. The limitations of the model are its lack of feedback

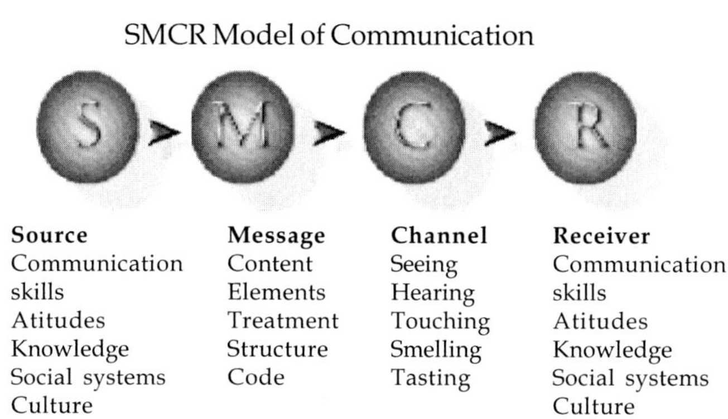

Source: From the Internet

Terms used in the chapter

Mass-media

It is a collective phrase that represents not only the press, cinema, radio, television and internet, but also to some extent, books magazines, pamphlets, direct mail literature, posters, folk media, and natural communication methods such as rumours, education and preaching. It is so termed because its reach extends to vast heterogeneous populations. Generally the mass media employ technological means to communicate to the masses. They are founded on the idea of mass production and distribution. Wiebe defined mass media as those readily available to the general public.

Selective attention

The media are full of competing messages. The process of screening vast amount of information in which one has no interest through mental filters is called selective attention, for example, an adult will be more tuned to listening to the news while a child would rather watch a cartoon show.

Selective perception

This is the tendency to interpret communication messages in terms of one"s existing attitudes. People of distinct psychological character same media content in different ways. This depends on factors such as age, values, family, opinions etc. Selective perception is influenced by social relationships.

Selective retention

The ability of an individual to retain certain messages in his mind while ignoring others is called selective retention. This is influenced by various psychological and physiological factors such as choice, values, culture, emotions etc.

Selective exposure

Some individuals are exposed to certain media effects/messages while some are not. This screening aspect depends on many factors such as reach of media, accessibility, age, cultural acceptability, taboos, etc.

Opinion leaders/change agents

The opinions of people in a group are influenced by what they hear from "opinion leaders". An individual who is a member of a group manifests certain characteristics in his thinking and behaviour that contribute to the formation of "public opinion". The opinion of the leader is based on rational thinking due to education and experience. They weigh the pros and cons of the information they receive and then give their judgement on it.

Encoder

In the process of communication, the sender or source of the message is referred to as the encoder.

Decoder

The person receiving the message and decodes it is referred to as the decoder.

Feedback

Feedback, term form cybernetics, the study of messages. It refers to an inquiry, response or experiment. Feedback can be positive (when the required result is achieved) or negative; instantaneous (when the response is immediate) or delayed. Feedback is used to gauge the effectivenss of a particular message put forth or situation that has taken place.

Noise

In all communication, there is a sender, a message/ communication and a receiver. The meaning of a message is greatly dependent on the culture in which it is transmitted. The sender encodes a message, the receiver decodes it. Between the sender, the message and receiver, noise gets in the way and complicates the process. A noiseless communication does not exist. There always is some kind of noise entering the communication. Noise can be physical noise for example static or psychological i.e. when culture, taboos or values come into play to disrupt the normal transmission process of communication. Misunderstanding of a particular message i.e. distortion of meaning is a form of noise, example, the game of Chinese Whisper"a person starts off with a particular message and the original message may be distorted by the time it comes to the final player.

❑❑❑

Chapter 4

Dimensions of Journalism

Introduction

Mass media can be divided into printed, spoken, visual, and a combination of those three. Mass media materials are needed whenever large number of people must be reached and motivatedto action, which is an important advantage. Printed media, radio, television, cinema are examples of

Mass media, which have large audiences with very low cost per person contacted. Modern nationsprobably could not exist without mass media materials.

Print Media

Print Media of various types such as leaflets, pamphlets, circular letters etc. Printed materialsare used singly or in combination with other extension methods. Some of the printed materials arepresented below. Print media, Radio Journalism, T.V. Journalism,Photo Journalism, Audience Research.

Leaflets

A leaflet is usually a single sheet of printed matter, sometimes folded. It gives accurate orspecific information on a particular topic.It is intended to:

1. Provide precise and reliable scientific information told in simple language about a singlepractice or item of interest.
2. Serve immediate needs of the farmer.How to write

A. Plan the Script

1. Decide on your message select topics related to the urgent needs of the farmer - have onesimple practice or idea at a time.
2. Jot down all appropriate points - decide the essential points - finally arrive at the desirablepoints.
3. Select the most important one from the essential points - this forms the central theme.
4. List the remaining essential points in logical order and group the desirable points under theappropriate essential points.

B. Write the Script

1. Write your script, with all essential points in sequence, the desirable points supporting theessential ones.
2. Make the most important points catch the reader' interest.
3. Write in simple short sentences - use familiar words - be clear in your words and sentences avoid being misunderstood.
4. Address your sentences to your reader.
5. Lead him to action.
6. Be accurate in your information.
7. Be brief.
8. Use illustrations and pictures in the appropriate place.
9. Give details with reference to local situation.
10. Start with an appeal, which will be of benefit to the reader.
11. Round up with confirming what you said at the opening of your leaflet. It will reassure your reader.

12. Mention the source where further information on the topic or help can be obtained.

C. Review the Script

1. Go over the writing, after the completion of it – over after a day.

2. Remove the defect and rewrite, where required.

Leaflets have several advantages. They are (1) economical, (2) relatively easily and quickly prepared, (3) can be preserved and used by readers, and (4) supplement other information and media. They can be used in schools, in farmer training programs and as a basis for discussion in meetings. Commercial concerns and theatres commonly distribute leaflets to advertise their products. In extension, leaflets may be distributed by the staff personally at field days, farmers' tours and collegedays. Other distribution methods are by post, and by placing in seed, fertilizer, and insecticide bags.Such methods permit large numbers of leaflets to be distributed easily, quickly and at small cost. Attractiveness of leaflets can be increased by using different colors of paper or ink, and by illustrations or photographs. Leaflets - sometimes called flyers.Here are a few examples of leaflet writing: Take Care of AilmentsEven when a few days old, calves can get a number of disorders. You cn save them from death if you start treating them early.Calves of all ages get 'white scour', caused by germs. This is a digestive trouble, and brings in diarrhoea. Dirty conditions in the shed, not feeding the calf with colostrum and want of vitamin A in the feed, all lead to white scour. Print media, Radio Journalism, T.V. Journalism,A calf with white scour gets fever, shows dullness, restlessness, and a dislike for food. Shortly, it starts purging a yellowish white liquid, which gives out a bad smell. Very soon, the calf becomes weak and is unable to stand. If the calf is quite young, it may die.Diphtheria is another calf ailment. Germs entering the calf's mouth cause lesions. The calfgets fever refuses food and shows dullness. The inner lining of the mouth will be seen swollen. Saliva starts dripping from the mouth, the calf coughs and gives out a dirty yellowishdischarge from the nostrils.

When the disease is severe, calves die in about four days. 'Navel ill' or 'joint illness' is yet another serious ailment of calves.

In this, the navel gets swollen, and gives out a foul-smelling discharge.

You may also notice fever, weakness and dullness in the calf. The joints may also get swollen,hot and painful. Treat all these cases early. The veterinary doctor will tell you what treatment to follow. Prevention, however, is always better than cure. For this, you had better follow these practices.Provide the calf with a clean, ventilated house. Do not overcrowd calves in one place. Feed the cow properly before claving. Treat the calf's navel with tincture of idoine soon after birth. Feed the calf often with small quantities of feeds; never overload it with food. Give a spoonful of cod-liver oil to the calf each day for the first four or five days. Exercise the calf daily. Also, when you notice a calf suffering from any of the ailments: Keep the sick calf away from the healthy ones Use separate feeding utensils if possible. Get expert help from the veterinary doctor. Better Level and Bund your Field

- Is your field sloping or uneven? Then it is time you leveled it up properly.
- If your field is sloping or uneven.
- A heavy rain will wash away the good soil at the top.
- The rainwater will run over the field instead of soaking into it. The crop, therefore, will notget enough water to grow well.
- You cannot sow the seed evenly. In some places it will be covered deep, in other places itwill be left on the surface.
- The seed will not germinate well or uniformly. A poor germination means a poor crop.
- You cannot inter-cultivate your crop easily.
- You will need to give more irrigation to your crop.
- Much of the water will be wasted, as it will run over the field, instead of soaking near theroots of the crop plants.But if you level your field...
- You will save your soil.
- You will be able take the best use of rainwater.
- You will spend less on labour for farm operations.
- Your crops will grow better and yield better.

You can level your field...

- By digging, carrying and dumping the soil, by using a buck-scraper or a wooden levelingboard.To level your field easily and cheaply, use a buckscraper. It scrapes the soil, carries it and dumps it where you want it to be dumped. Lift the handles up a little, and the buck scraper scrapes and collects the loose soil. Let go the handles, and the scraper carries the soil where you want it to be carried. Lift the handles andtilt the scraper, and it dumps the

Soil. Print media, Radio Journalism, T.V. Journalism, Photo Journalism, It is better to plough the land before you work the scraper. You can use the scraper to erect bunds also.

Pamphlets

The leaflet contain from 4 to 12 pages stitched or stapled at the center. The pamphlet varies in size from 12 to 24 pages. The difference is arbitrary, and may sometimes refer to a leaflet as a pamphlet or it may be the other way round.

Circular Letters

Circular letters give an intimate, personal approach to the extension message. Receiving a letter is an important event for village people. Use letters as an effective method to convey information of common interest to large numbers of literate people at one time Circular letters are more easily and quickly prepared than most other information materials. They are economical when mimeographed, costing only a fraction of paise each. It can be very effective if it has a personal touch and drawings, cartoons or pictures.

There are two types of circular letters: (1) Announcement and (2) Subject Matter. The Announcement letter announces about an event such as meeting, demonstration etc., to take place. It invites people to take action. The Subject Matter type of letter may be in the form of a chatty, informal news story. It has the facts, but they are presented in personalized form. It is designed to bring out changes in practices. A good circular letter contains these three essential parts: First - introduction. Relate the new message to a previous favourable programme, experience or situation, or show how it is opposite from a previous unfavourable programme, experience or situation. Second - explain the importance of the topic and show how it will benefit the reader. Third - appeal to action and relate the action to the objective in writing a letter.

1. Have a definite reason for writing a circular letter.
2. Decide on the main idea, the length, the tone, and the illustration before you begin writing.
3. Suit your salutation to your audience.
4. Pack your opening sentence with the main idea plus reader pull.
5. Keep your sentence short, full of personal references, and free of hard words.
6. Keep your paragraphs down to 80 words.
7. Choose your illustration carefully and execute it well.
8. Tie in your closing with your opening, ending on an action note.
9. Give your letter "sales appeal" with a pleasing appearance.
10. As far as possible don't write more than one-page letters.
11. Use coloured paper at times to add atmosphere.
12. Wide and even margins in a letter contribute to the ease of reading.
13. Too much detail will get poor response.
14. Give simple, accurate, complete information and direction.

- Circular letters can be circulated by:
- Direct mailing to all concerned,
- Distribution through village institutions like Panchayat etc., and
- Sent periodically or on special occasions.
- Here are two examples how circular letters are written
- Better Farms Make the Fallow Work

Dear Friend

In the next few days, you will be harvesting your paddy crop, and perhaps wondering if there is some way of making use of the land instead of allowing it to lie fallow till the next paddy crop is planted. Many people with the same problem have solved it by raising a crop of groundnut. This is giving them ready cash. Your co-operative society says that it can sell the entire groundnut you can produce at a good price the very day you bring it in. The market is up. I am sure you will be interested in trying groundnut this year. You know groundnut is a legume; it will therefore leave your land richer (in fertility) when you take it off the field. You will also have plenty of green matter to plough in. Many have found that this is a good green manure for their next paddy corp.You can get the groundnut seed (ask for the variety TMV-5) from your agricultural store or the co-operative. When you get it, just give one ploughing to the field and sow the seed in lines with a spacing of six inches between seeds and nine inches between rows. Get enough seed to sow all the fallow. Calculate the seed you need at 90 pounds of kernels per acre. You will have to pay only 25 paise per pound of kernels, but just think of the profit it will bring you, if you have any difficulty in getting seed for sowing, I shall be able to help. Call at my office next time you pass this way. I will be glad to meet you. Remember, you can put that fallow field to work

- You're sincerely,
- Rama Murthy
- Extension Officer
- Want your vegetables to be fresh?
- Dear Housewife,

Surely, you want your vegetables fresh, even in the middle of summer. You must be having difficulty in keeping vegetables fresh in the kitchen in this weather. The heat makes the vegetables wilt and dry up so badly that you probably have to throw out some badly wilted ones every day. How irritating.

But there is one easy way out. You just put the vegetables in alkathene bags. They stay fresh for six days! The Home Science Wing at Gangavati tried this method last summer, and found it to be a verygood one for house wives like you! You can buy these p1astic' bags in your village shop. They cost three paise each, or, if you look into your box you will find some, which you use for wrapping clothes or saris. These will do. Keep some bags handy, and wrap the different vegetables in separate bags. Of course, wipe the vegetables well before you put them in. Press out all the air from the bag, and tie up the mouth tightly with a thread. Take out the vegetables whenever you want them, but tie up the bags again in the same way. The only thing you have to bear in mind is that you do not leave any air in the bag. I am sure you would like to try out this method. You will see for yourself how your vegetables will remain garden-fresh even after five days.

Sincerely Kumar Mukhya Sevika

News Letters

A newsletter deals with current items of interest and normally is printed periodically. Newsletters deal with more than one subject, are printed in large numbers and distributed in manyways.

Some tips on preparing news letters are:

1. Use a salutation, if necessary, such as "Dear Farmer" etc.
2. Let the length of the letter be normally 4 to 8 pages.
3. Have desirable size like 6" x 9". It should be handy.
4. Illustrate it with realistic photos, cartoons and sketches.
5. Let the lay-out be attractive. It is better to have news items in columns and each news item would carry one illustration along with it, as far as possible.Print media, Radio Journalism, T.V. Journalism.
6. Have an attractive get up. The front page has to carry title of the news letter, with an illustration containing emblem – representing the publishing agency. The page may contain a brief editorial. Last page may contain again an illustration, special appeals, if any, and name of publishing agency.
7. News items may consist of periodical hints to farmers, achievements and experiences off armers, research findings, announcements, and a few statistics. Other miscellaneous items may be important events of the area, special festivals in which the audience are interested.

Folders

A folder is a single piece of paper folded once or twice. When opened, material is presented in sequence. Make sure this sequence appears in the finished folder. If not, the reader may become confused. Folders are normally printed on paper heavier than flyers so they may last longer. Folders are usually prepared on "offset" or heavier paper. They may be made more attractive by using photographs, line drawings and various colors of inks and paper. A four-inch by eight-inch folder is quite attractive. A width-to-length ratio 1: 11/2 may be more suitable when paper size permits, without waste. There is no set rule on size. The basic consideration is that publication size fits the paper stock, thus eliminating excessive trimming. Folders are economical. Folders are not distributed as freely and indiscriminately as are flyers because they (1) cost more, (2) may be used with more sophisticated audience, and (3) are moredurable.

Bulletins

A Bulletin conveys large amount of information. Its primary objective is to give information which the reader can apply to his own local situation. It is a booklet running into more than 20 pages.

Bulletins are of two kinds

1. Technical bulletin - designed primarily to present scientifie material to those working in specific fields.
2. Popular bulletin - to present material to people in the field of extension.

Extension bulletins are popular bulletins for extension workers, progressive farmers, school teachers, instructors in colleges and extension training centers. For effective preparation of popular bulletin the following points have to be considered:

1. Have good 'eye appeal'. It should be attractive to see.
2. Use direct style in writing: It should be simple, direct, clear and convincing without 'fine'writing.
3. Choose words as you would your friends. Select words and use them in proper relationship.
4. Make sentences light. Be clear, precise and definite. Make sure you say what you have in mind. Check and recheck. Mention important information in the first part of your sentence.
5. Have short paragraphs.

Farm Journals

Farm Journals are important tools to disseminate relevant development messages to the rural audience. The Farm Journals suffer from a number of problems viz financial problems, delayed publication, lack of trained farm journalists, technical jargon etc. Some of the suggestions for improvement of farm journals are.

- Farm Journals must cater to miscellaneous interests of farmers and provide contents to tastes of different categories of readers.
- Use of illustrations and photographs should be promoted.
- Editor of farm Journals should preferably be agricultural Graduate trained in farm journalism and exposed to rural life and farming.
- There is urgent need to simplify technical terms.
- Lay out, typography, preparation of illustration, proof reading should not be left to chance.
- Accuracy in reporting of technical information is a must. Name of chemicals, varieties and their doses must be carefully checked.

News Stories

News is any timely information that interests a number of persons, and the best news is that which has the greatest interest for the greatest number. It is an accurate, unbiased account of the main facts of a currant event that is of interest to the readers of newspaper. The purposes of new stories are a) to develop interest b) to inform general public c) to disseminate subject matter information d) to create favorable attitude e) to reinforce other extension methods like meetings and demonstrations.

Remember that news should have one or more of the following characteristics :

a. Something that actually happens
b. Unusual (extra-ordinary)
c. Important (not trivial)
d. New, recent or timely
e. Something that interests farmers
f. New knowledge

Use accepted principles of good writing :

i. Write the lead sentence i.e. something of importance to the reader
ii. Use inverted pyramid pattern of writing

iii. Use the five W's and H as a guide
iv. Write in simple language
v. Avoid using your personal opinion
vi. Be accurate, fair and brief
vii. Include motivating appeals
viii. Evaluate effectiveness of news stories

A few examples will make this clear

Soils Need fertilizers

New Delhi August 7- About 18.4 million tones of nutrients are removed from the soil every year by the different crops in the country according to agricultural experts.Writing in the handbook of agriculture brought out by the Indian council of Agricultural Research, the experts point out that in terms of fertilizers, the removal rate is 4.2 million tonnes of nitrogen, 2.1 million tonnes of phosphate, 7.3 million tonnes of potash and 4.8 million tonnes of lime. Production of larger yields through improved varieties of crop and intensive cultivation will increase nutrient removal further and they warn that unless the soil is given sufficient supplies of chemical fertilizer and other nutrients, the drain in manurial supplies will continue to make the soils poorer. Feeding chicks the earlier the start the better. New Delhi: January, 20 – with holding food from chickens for the first 48 hours of their life is not a sound practice. In fact, poultry experts here point out, the sooner they begin feeding, the better is the start they get in life.Chickens without food sometimes start pecking at each other, some pick up litter, which cause internal disorders.

Chickens should also have, the experts say, enough hopper space. Other wise the stronger ones push out the weaker ones, which develop poorly. Feather pecking and cannibalism in chicks are also traced to lack of enough hopper space, according to the experts.

Wall Newspaper

As the name implies, a wall newspaper is printed on only one side so it may be displayed on walls or similar backgrounds. They are intended to provide timely information periodically to rural people. Color, illustrations, drawings and photographs help attract attention. They are normally on off set paper. Unlike leaflets or folders, wall newspapers contain more than one item.

Announcements, crop forecasts, latest research results, success stories, recommended practices news of extra activities and other important items, all may be in one issue. Some points to remember in preparing and using wall newspapers are: Lettering should be bold and large enough to be read from distance, words and sentences and paragraphs should be short, select items for interest and timeliness, display in prominent, well lighted open areas at convenient height to reading, and protect from sun, wind and rain when displayed outside.

Newspaper

It is a bunch of loose printed papers properly folded, which contains news, views, advertisements etc. and is offered for sale at regular intervals, particularly daily or weekly. Newspapers are usually printed on a special type paper known as newsprint. Figure 3 Wall Newspapers Print media.

The purposes of newspapers are

- To serve as a forum for extension activity in an area or a community
- To put out information which would be of service to the leaders and people and to highlight the important activities of individuals and groups which are worthy of emulation.
- To acquaint the public about programmes, activities, progress etc.

The points to be remembered in writing newspapers are a) write it in story type. It should be easy to understand and simple language b) give a personal touch c) give accurate details d) plan publication so that the different issues are in a sequence e) encourage local farmers to write and f) givea suitable caption.

Banners

Banners are traditional mass media information visuals in India. One need only walk down a village, rural town to see banners flying, advertising every thing from meetings to sweet meat shops. Extension workers seldom use them.Banners are rather costly and require considerable time and skill. As with all other information materials, they must be attractive with a brief, clearly presented message. Height and length of a banner should be pleasing proportions unless durable cloth and water proof paints are used. Wind and rain may destroy the banner. Therefore flex printing can be taken.

Calendars

Calendars are widely distributed in most countries of the world as an advertising tool. Hence they should not be overlooked by extension workers. A unique and outstanding advantage of calendars, when hung in offices, colleges, schools and homes, is that they are constant, year-round reminders of educational messages.

Radio Journalism

Radio is most popular mass media, among majority of rural people. When you want to reach people who cannot read or write, or people who live in remote villages, and when you want to reach people speedily, you make use of the radio. It is a personal medium, received in private by the listener in the company of his family members or by himself. The radio belongs to the spoken word means of Communication. The medium is transient and fleeting and its impression is quick and faint. It is a one-way communication. The message has to be simple and clear so that people can understand it and act. The broad caster has to get and hold the attention of audience otherwise the message is lost. Special features of Radio: There are certain special advantages with radio as a medium. It reaches more people morequickly than any other means of mass communication. It is relatively cheap. It reaches illiterate audience also. It builds enthusiasm and maintains interest. You use the radio to inform alert, direct, interest, stimulate and motivate people.

Limitations

- Radio alone cannot bring change
- Listeners have no control over content or speed of message
- It is difficult to transmit complicated messages through radio.

Mode of programme presentation Radio programmes are presented in the form of news, interviews, discussion, documentaries, quiz, question-answer etc.

Writing the News

Let's take radio news. Writing the news for the radio is different from writing it for the newspaper. The technique you follow for writing the news for the newspaper is good only for the news paper. The reader has your words before him and he can read them at his pace. There is the headline to prepare him for the news. He can go back and read it all over again if he misses any point or fails to understand you fully. Not so with the radio. There is no chance for him to go back and start from the beginning.That is why you have to write the news for the radio to suit the ear and not the eye. It has to be short, simple and direct with familiar words and simple sentences. In other words, it has to be made very, very simple Write it the way you would say it aloud to your friend or a group of peopl. Remember, it is a voice-to-ear method of communication. Your news script has to be 'listenable' as your written word has to be 'readable'. Let us take a few examples. Readable: A bullock–drawn reaper for harvesting wheat has been designed and developed by the Research and Training Center of the Punjab Agricultural University at Ludhiana. The machine has been successfully tried on the University Farm, and steps are being taken for its production on a commercial Scale. Print media, Radio Journalism, T.V. Journalism.

Listenable: A bullock-drawn reaper has just been developed. The machine is for harvesting wheat. The reaper has been designed and developed at the Punjab Agricultural University at Ludhiana. The machine has been worked successfully at the University Farm. Steps are now being taken to produce it in large numbers for sale to farmers. Readable: The Uttar Pradesh Government today proposed the establishment of four new fertilizer plants with a capacity of 1.8 million tons a year to meet the State's requirements of fertilizers in the Fourth plan period. Listenable: The Uttar Pradesh Government today proposed to set up four fertilizers factories. The four factories together will produce 1.8 million tons of fertilizers a year. This quantity of fertilizers is what is needed by the state in the Fourth plan period. Thus you see that writing the news has to be based on sound rather than sight. It follows the rule of good conversation. It is more direct, with a simpler sentence structure.

Keep the news items brief. The radio station cannot stretch time, and has to accommodate your programme within the time slot it has at its disposal. Script for the talk A straight talk on the radio is in a way the most unnatural way of talking. It is an indirect talk with an unseen listener. It is a one-way traffic. Therefore local dialect must be used like bhojpuri, dharwad kannada, telangana, coastal telugu, rayalaseema telugu etc. Thousands may be tuned in to your station and to your programme. But you have to addressyour talk to one, two or to a small group of listeners. You have to mould your talk as if you are sitting round the table with the listeners and chatting, and not mark the words, teaching, lecturing or sermonizing. You have to remember that the listener is free to turn the knob if you sound as if you are teaching or lecturing. Nobody tunes in to hear a dry lecture or a sermon. First, select a topic. It should not be one that needs an hour to explain. It must be something that you can cover in a few minutes. It has necessarily to be of interest to a very large number oflisteners.You have plenty of topics to select from, such as significant crop or livestock developments, the agricultural situation, demonstrations, success stories, current problems, research station stories,meetings held, or to beheld, questions asked, new farm or home publications, current angles of aproject, crop and livestock competitions, activities of farmers clubs, women's clubs and youth clubsand how-to-do information.A five-minute talk is ideal for the radio.

A talk should never go beyond ten minutes at anycost. People speak at different speeds, but 120 words should be a good average for a minute. At thatrate, your talk should have about 600 words. You can have an additional 50 or 100 words as a standby,in case you finish reading earlier than your allotted time. Since you have a limited time, select only one phase of the subject or topic you will like totake up. Write out the central fact or point as a complete and definite guiding statement before composing your talk. This should be such that listeners would be able to remember clearly and easily. Select two or three supporting points, which will strengthen the main statement.

Then, write the script. When you write, write as if you are talking to a person in his own home, someone whom you know personally. Make your script clear and convincing, and your arguments logical so that it stimulates and maintains enough interest in the listener to make him seek more information of the subject or take further action. Unlike other forms of writing, the script for the radio talk should have your idea plainly stated at the beginning itself so that the listener has no doubt at all of what you are going to talk about. Having done that, you enlarge on the main idea - provide the supporting ideas.

Finally, summarizeclearly what you said. Whenever you want to make an important statement, alert the reader in advance so that he may not miss it. Say, "Now you will see how this is done", How do you apply this chemical?". "You must know how this food nutrient is lost", "It is important that you know how this is made". Or "I will now tell you why this is so". Make your facts and statements convincing. Give logical reasons for making them. Give examples. Quote authorities. Give instances when what you say was put into practice with success. Print media, Radio Journalism, T.V. Journalism.

Point out the results of experiments, trails, demonstrations and research surveys. Give local places, names of local people, and local examples. Spell out large figures in the script; write two lakhs, rather than 2,00,000. It makes reading itout easier for you. Avoid giving specific numbers; round them up. Nobody will remember "24,858 hectares" but "about 25,000 hectares" is easy enough to remember. Do not give the impression that everything to be said about your subject has been said in yourtalk. Make the listener seek further information about it, either by contacting the specialist or askingfor a leaflet. Providing all the information on a subject is not the job of the radio. After you have written the script, check it. sSee whether you have presented the subject correctly, and whether you have done it clearlyand briefly. See that all the words are short, simple and easy to pronounce or listen to. Check and seethat your sentences are short, and the sequence logical. See if you have included human interest, andif the talk can leave a message behind.Read the script aloud. See if you sound, as you should, as if you are talking to some one.Then write the script on a soft, non-crackling paper. Provide a broad margin. Use plenty ofspace between lines. Indent your paragraphs properly. That is, start it quite away from the margin, so that before you come to it you know you have a new paragraph. Do not carry a part of a sentence on the next page. Other wise listeners will hear a pausesomewhere in the middle of the sentence. Correct your script carefully for mistakes, and mark theplaces where you want to give a pause like this: When you want to emphasize a word, underline it. Mark your pages, and put them in properorder.Another important aspect is that the script has to be divided into different parts depending upon the nature of the message to be conveyed. The following parts may constitute an effective scriptdealing with an improved agricultural practice.

1. The first part should be designed to attract the attention of the listeners towards the subject proper. A strong opening makes people want to listen.

2. The second part may be analysing the present situation laying special emphasis on the problems encountered based on the local needs.

3. The third part may give out facts about the recommended practice and its merits over the previous practices and try to win the confidence of the listener.

4. The fourth section may deal with an appeal to action.

5. Finally the script may end with a summarization of all the different parts. This will give an opportunity for listener to know all-important ideas in the talk and get as reinforcement.

The following is an example of a well-written radio script.

For some years past, many small farmers in Saran District have been growing coriander in sugarcane fields. They have a very good reason for doing so.

I shall talk to you today on the growing of coriander in between sugarcane rows.

"You will surely ask, first, why you should grow anything at all in a field of sugarcane. The answer is, if you do so, you get some extra money. And you get this extra money without bringing down the sugarcane yield or quality in any way.You will then naturally ask me, why coriander? Why not mustard, linseed, cumin or some other crop? Well. We tried all these crops at the Agricultural Research Station at Sepaya in Bihar. We tried them for years, to make sure which is the best for growing with sugarcane. We found coriander the most profitable one. When coriander was grown along with sugarcane on one acre of land, it gave more than two hundred rupees as profit. The other crops gave much less. Mustard was the next best. Mustard gave one hundred and ninety rupees an acre. It would have been as good as coriander, but it created other problems. It began spreading over the young sugarcane plants. The cane plants, therefore, tillered late. Mustard plants later lodged also, thus making it difficult to work implements for inters cultivation. That is why I am recommending that you grow coriander.You may now be thinking: why, will not the yield of sugarcane go down because of the other crop in the field? The answer is no. We kept a watch on the sugarcane. From time to time, we tested the sugarcane to see if its sugar content was affected. But, no coriander in no way affected the sugarcane crop.Print media, Radio Journalism, T.V. Journalism, I shall now tell you how you can grow coriander in the sugarcane field. After you have prepared the field for sugarcane, you open up furrows three feet apart, and plant the cane setts as usual. You cover the setts with a four-lined cultivator, and follow it up with a henga. Then pass a country plough between the cane rows and sow coriander. Work the henga again. You need about five seers of the seed to sow an acre this way. One word of caution. Do not delay sowing. The earlier you plant sugarcane and sow coriander, the higher is your profit. Next; you would like to know how much of care you have to give to coriander. I assure you that no extra care is needed for coriander at all after you have sown it. Whatever operation you do for sugarcane is good enough for coriander too. Of course, if there is no rain in January, you have to give irrigation both for the cane and for coriander. If you sow coriander, say in October, it would be ready for harvest by the end of February or beginning of March. Of course you know that you have to pull out the plants and dry them. Thresh the seeds and dry them in the sun. Winnow the coriander and store or sell. You can harvest about seven mounds of coriander from an acre. After removing coriander from the field, give a deep inter cultivation to sugarcane. A fort night later, irrigate the crop. This is all that you have to do for getting a crop of coriander as an intercrop with sugarcane. It is so easy to grow and I am sure you will also like to try it on your field. You can try this crop without any fear. The many small farmers in Satan District who have tried it find the side-income helps them meet the expenses on the sugarcane crop. Remember, sugarcane is on the land for twelve months or more you have to spend a pretty high amount of money on its cultivation, and have to wait for a long time to get the money back. But when you grow coriander along with it in between the rows, you can get back a part ofwhat you have spend in a few month's time. Try coriander as an inter-crop when you plant your sugarcane next time, and find out for yourself how will it pays you.

Television Journalism

Television has unique advantages over other mass media. While it provides sound, vision and movement. It can reach largest number of people in shortest possible time. It is a medium unlike any other. T.V. Screen is small and depends on close-ups. People can watch sitting in their homes. Television viewing dos not demand extra strains to go out or read a book and the message are pre-selected, sorted out to present in simple manner. The medium is quite suitable for subjects that require dramatized presentation, identification of objects, live demonstration of complex technical process depicting animated presentation and presentation of experiences, places, and processes, unfamiliar to viewers.What television can or cannot do as a mass medium T.V. leads to awareness, contributes to information and helps to form opinion. Before the farmers are motivated to action, televised information may create awareness and interest. However, local demonstration and interpersonal communication are required to convince them. T.V. is good in exposing people to a whole range of new ideas and experiences. Programme on agriculture have immediate effect if ideas are put forward at a time when need they most, deal with subject about which farmers do not have pre-fixed notion. If programmes are supportive to efforts made by extension or advisory services working in the locality, the changes are more likely to come.

Limitations

It is centralized mass medium i.e. programmes are produced in distant urban environment and so may not be appropriate for all viewers alike. Those who cannot afford costly inputs recommended by TV cannot benefit. Language of TV is quite sophisticated for rural viewers.

Components of good television programs. A good television program must attract the viewers' attention and hold their attention. It will do this by being-

Relevant to the audience

- Understandable
- Believable
- Technically well made

Characteristics of effective television program

- *Good Design*: Purpose, objectives and message will be strong and clear throughout and will be based on thorough planning. There will be an effective balance of visual and aural presentation of the message so that the visuals are not used as decorative additions to the spoken message.
- *Good Script*: The message will be presented in a format suited to the medium and attractive to the chosen audience. At all times the script will acknowledge the importance of findingand maintaining the balance between message and medium.
- *Good Production*: The finished program will be good, cleanly produced television, adhering to all the rules of good presentation — both technically and artistically. Attracting the attention of audience. It is good idea to give some thought to what it is that attracts audience to a subject with which they are not already familiar. Some considerations are:
- It affects their lives

- It affects their health
- It affects the welfare of their families
- It offers an immediate (or near immediate) chance of improved income of higher value formoney
- It offers an immediate (or near immediate) chance of improved life-style
- It allows them to make use of natural talent
- It provides them with a feeling of self-worth
- It gives pleasure
- It is informative.

In choosing format for the programmes, one should take into account the following:

- Audience preference
- Budget – some types of programmes are much expensive to purchase than others, so budget must be a high priority item in determining format. It is possible to make low cost programmes with help of Mass Communication departments of Education Media Research Centre of Universities.
- Time – this is as important as budget, since some TV formats such as animation, requireconsiderable time which may not be available to you.
- Type of information to be disseminated
- Preference of TV station using the programme
- Length of programme (some formats are suited to longer programmes, while others are better for short "*filler*" style).
- Availability of facilities (since producing television requires an array of facilities including editing, dubbing, transfer time etc). One must make sure these facilities will be available when needed or even that they are available at all.
- Producer's personal preference and abilities
- Suitability of format to subject matter

Formats commonly used for the delivery of development of message on television:

1. *Actuality*: In actuality, the audience sees and hears people in undertaking being discussed.

   The programme may be introduced by a host, but from then on, rather than interviewing people the programme cuts directly to the people involved.

2. *Animation*: It is business of combining individual shots (of still drawing) to give the illusionof movement.

3. *Demonstration*: A demonstration programme is just what its name implies. It is a programme in which the main focus is the demonstration of the way something is done or the way something works. A demonstration can be done in the studio or in the field.

4. *Documentary*: A documentary explores a topic in depth. Frequently a documentary is created by shooting first and scripting later.

5. *Drama*: A drama is a play which presents a *true-to-life* story in dramatized form with actors playing the parts of the story's characters. The story develops through what the characters do and say. Drama is an effective way of involving an audience since it is a realistic representation of life.

6. *Graphics*: The word graphic means pictures. A very effective programme can be created using painted pictures, across which the camera can move to create a sense of movement.

7. *Illustrated Talk*: In this type of programme, the focus is on person (usually an expert, orperhaps a host) who is presenting the information. However, rather than stay on the person (and create a boring type of programme known in the trade as a "talking head" programme)the programme uses illustrations of what the presenter is talking about.

8. *Interview*: This is exactly what the name implies – a host interviewing someone on a particular topic. An interview can be done on the location or studio. The interview is a very hard programme to do well on television.

9. *Panel show*: This type of programme is usually done in studio with a group of experts discussing a topic with a moderator. It is likely to be more interesting than an interview because it involves

more number of people, but it needs to be very well planned and rehearsed so that the camera is always on the appropriate speaker.

10. *Puppets*: Some very effective and interesting programmes can be made using puppets instead of live actors. The advantages of puppets are that they add novelty; they do not have to be paid (as actors often do); lines do not have to be learned, since the people playing the parts of the puppets can read the lines and rehearsal time can be considerably shortened.
11. *Slides*: Slides are much cheaper than film or tape and can be used very effectively for a television show. Be sure that one allows about 1 slide for every 3-5 seconds. If the slide is left on screen for too long, it slows down the action very badly and results in static and therefore boring show.

Writing Script for Farm Telecast Programmes Script

Writing for farm telecast is the most difficult job. It requires:

1. Great degree of creativity
2. Language and editing skills
3. Journalistic talents
4. Command over the content
5. Capacity to put the content in a story format in logical sequence and suitable format
6. Perfect visualization of entire programme in the mind
7. Selection of correct format (method of presentation)
8. Profile of audience
9. Knowledge about local terminology and illustrations
10. Information requirements of the farmers or viewers

Content outline

The main purpose of preparing an outline of the content is to explore important ideas for writing a video script. It is the first attempt of a writer for making a video programme. An outline should be:

1. Not so detailed
2. Not very short also
3. Should contain important information of the subject intended for the clients. A good outline will state:
   - Title of the programme
   - Its aim and objective
   - Approximate duration
   - Sequence of events
   - Two or three times treatment of subject in different ways for making different approaches tothe topic for better organization and presentation of video programme in an appropriate format. Print media, Radio Journalism, T.V. Journalism, Writing a Story BoardAfter delivering a story, visualize the situations you are going to describe. Story board meansnot only asking of words but learn to think in pictures and support with narration visualization by making simple sketches or still pictures of content in sequence with brief narration, in allocated time and suitable exposure such as close, long shot or medium one depending on the importance ofpicture. A story board can be prepared on sheets of plain papers or in a printed proforma which isenclosed herewith. After arranging the outline and story board, write your detailed blue print of programme. This script will be a guide map for the definite direction for shooting works, art work, audio recording and video recording. A script is a sequence of subject matter of sequence of subject matter of pictures description with narration. The placement of the camera for each scene will be indicated along withtime segment, background music, etc. Print media, Radio Journalism, T.V. Journalism, Photo.
     - Exposure of subject of long distance of picture is a long shot.
     - When lens covers only the subject nothing else is a medium shot.
     - Concentration on subject is close-up shot

They are expressed as L.S., M.S. and C.U. While preparing a script for video production, one should keep in mind the following points for an interesting and effective programme production.

a. Capture the attention of the viewers with a dramatic sequence illustrating the importance of the topic, with good music, scene and correct informative opening.
b. Use local works and illustrations
c. Analyse the present situation in your narration and introduce new innovation.
d. Create interest and humor in your script
e. Consider the value of music as a complement to narration
f. Summarise the salient points.

Giving Commentary for Farm Telecast

VHF Communication

Visual: Memorize pictures, images, diagrames, charts, graphics, etc.

Hearing: Memorize sounds, conversations, melodies, accent, etc.

Feeling: Memorize emotions, smells, tastes, tactile experiences, pain, and joyA good video production provides a wide range of stimuli for its varied audience. Commentary is one of the possible channel as it enable audience to memorize picture with music, with words by3 ways i.e., visual, hearing and feeling VHE communication.How to Write For Visual Verbal MediaWord choice- Use familiar, simple words Sentence length- Short sentences are important for spoken narration. Thumb rule: Sentence should have 8 to 10 words. Print media, Radio Journalism, T.V. Sentence structure- Variety in sentence structure captures viewers wondering attention. More personalized writing makes listening interesting. Ex: Real simulated conversation or dialogue gives personal touch. Sentence structure acts as mask to meaning. Have balance between simplicityand interests. Farmers, particularly tilling poor soils, can increase crop yields by using fertilizers. Farmers can increase crop yields by using fertilizer. For farmers tilling poor soils, fertilizing is must.Rules of thumb while writing for commentary

- Let picture tell as much story as possible. Use words to interpret the picture but not to describe it
- Do not tell about some thing not shown in the picture

- Do not write too much for each picture regardless of spoken or shown
- Do not vary frequently the pace of presentation for spoken narration max. Limit is 15-20 seconds per visual
- Use shorter sentences for spoken narration i.e., for commentary, write for intended audience not for literary critics
- KISS – Keep It Simple and Straight Use your voice

Projection : Projecting voice to the last row in the room

Articulation: Don't swallow words, beware of verbal tics

Modulation: Vary tone and pitch, be dramatic, confident and triumphant

Pronounciation: Beware of malapropos, watch accent

Enunciation: Over emphasize, accentuate syllables

Repetition: Use different vocal emphasis for key phrases

Speed: Fast delivery to excite and stimulate

Slow delivery to stimulate emphasizes, dramatize and control

Pause often – Silence is much longer for you than for audience.

Use humour- A laugh is worth a 1000 frowns.

Be enthusiastic- If you are not why should audience be ?

Take 3 breaths before you start. Never read anything except quotations

Music as Commentary music can be a commentary or spoken narration ?Yes, Example: Bhupala Rag played on flute Drawn in village when people start their daily chores Piece of folk music song to represent happiness of farmers Instrumental Veena music bit – while showing anything in action.

Photo Journalism

"One picture is worth a thousand words"

Words, written or spoken, often fail to convey what exactly is meant. Many words have more meanings than one, and the average reader, and especially our farm reader who has not had much education, takes time to understand the precise meaning of a word. The written word thus cannot tell everything clearly and precisely. This is where pictures come to your help.Man first created pictures to communicate with others. Pictures gave place in course of time to symbols. Symbols to letters and letters to words. Man found in words a quicker and easier form of expression, and words came to be used more than pictures. But pictures still have a hold on us. We are all interested in pictures, because they tell us more than what words can ever do. With pictures we learn faster, and remember things longer.A good picture (or illustration) draws the viewer's attention at once and arouses his interest in the Subject matter. Those who read slowly can more easily grasp meanings with the help of pictures. Pictures are a universal language, and help all readers get the meaning correctly. They also make the subject matter more realistic and lively.

Pictures help you cut down the length of your writing. A picture can replace the many words which you would other wise require to explain and convy a fact effectively. The average reader is always attracted to pictures and when there are pictures, and is readily coaxed into reading the text. Print media, Radio Journalism, T.V. Journalism, Photo Journalism, Audience.

The layout of a page looks more attractive with pictures. They also lend a tone of authority to your writing. All this makes Farm Editors not only give as many pictures as possible with the articles, but also sometimes set apart a few pages entirely for pictures either on different subjects or on a single subject to form a picture story. We shall learn more about the picture story later. Hence, when you write a story for farm papers or magazines, or draft a leaflet or bulletin, you think of illustrating it with photographs, drawings, charts and graphs. All these help a good deal in making your writing better understood. Of these, photographs are more commonly used, because they are more realistic than the others. You go in for photographs of people, farm animals, buildings and structures, machines,equipment, implements, crops, gardens, or outdoor and indoor scenes, depending upon the content of your story. Photographs are best taken by yourself if you have a camera and know how to take a goodpicture, since you are the best judge of the kind of photographs that should go with your story. If this is not possible, you can get them taken by a professional photographer. This, of course, will cost you more, and if the place where the photographs are to be taken is at some distance, your expenses on them will increase further. You can get photographs from other sources also. Agricultural departments, research institutionsand organizations, newspaper and magazine offices - all have collections of photographs, and you can get copies of the photographs you want from these place either free of charge or at nominal cost. A photograph must tell a story, and tell it at a glance. It should have one central or dominating point of interest. The other subordinate features should not take away the attention of the reader fromthe central point. This dominating point should not be at the dead center of the photograph, but a bit away from the center.

The photograph should be pleasing to the eye, if not fascinating. Light and shade will increase its worth. The horizon or skyline should not divide the picture horizontally into two; nor should the picture have a meaningless background. It should be clear in its details and have contrasts, which are needed for good reproduction.Professional photographers and Editors always talk of the 'composition' of a photograph. It means you include in the photograph what you want and exclude what you don't want. The best way of learning how to compose pictures is by observing how it has been done in a well-brought outmagazine.Before you finally select a photograph, see if it fits in with your writing, and if it helps clarifyand explain your subject matter, and can get the attention of the reader. Make glossy prints, of a suitable size (such as 8 inches x 10 inches, which is generally preferred by Editors). Discard all foggy pictures or those with muddy areas. Do not 'trim' or cut the photographs. Leave the trimming to the Editor. Do not also paste them on mounts. Type out (or write clearly) the caption and other instructions (such as acknowledgement, copyright, etc.) on a separate slip of paper and paste this slip on the back of each photograph. Get a thin sheet of paper to cover the picture or front side of the photograph and secure it firmly by pasting its top portion on the back of the photograph. This will prevent the photograph fromgetting scratched or damaged in any manner. Place the photographs between cardboards or any stiff material cut to the same size, and pack it for mailing. Do not use pins or clips. Do not write the captions or any other matter with pencil on the back of photograph. You are sure to ruin the picture. Never fold the photographs. They will break or crack at the folds, and the folds will appear when reproduced.You should know the mistakes generally seen in photographs sent for publication. The subject not being in focus is one Common defect. A blurred subject is another. Over-exposure and underexposure, spotting on the photographic print due to dirt particles on the lens, film or prints, damaged or scratched negatives, posed photographs of people specially dressed up for the occasion, unusual orun natural poses of people in unusual action or in unusual surroundings are other common defects,which you should avoid.

Drawings are also useful in explaining and illustrating the subject matter of a story. They suffer from the defect that they are not as realistic as photographs are but in certain cases they serve the purpose better than the latter. When the parts of implement or the working of a machine orequipment is to be explained, or when only one particular part of any subject to be stressed, drawings are more suitable. The drawings with pen and ink (Indian ink) sketch in such cases. Diagrams, floorplans, working drawings (blue-prints) are other illustrations are pressed into service to make our meaning clearer.

Charts and graphs are used for adding reader interest, and clarifying, condensing or explaining points. They also give feeling of authority to your writing. How to write captions. A good illustration speaks for itself and tells a story. Hence you need not explain it in so many words. However you can use captions (Also called 'cut-lines' or 'legends') to add to its meaning. You can also use them to generalize, modify or extend its meaning, to direct the reader's attention to aparticular point in the picture, or to identify the persons in it. When a picture is not related to the story or text appearing on the same page, ut is complete in itself with its caption. It is popularly called a 'stand-alone', and serves as filler. When an illustration will do well with an explanation give it as briefly as you can. The spirit of the picture decides the spirit of the caption. If it is light-hearted the caption may be so too. Also, the key details of the picture are the key details of the caption, and are normally given at the beginning of the caption. You must always give the caption to a picture in the present tense. This is because the pictureis 'frozen' at the moment. But if the reference is to some past action, then the caption given in the past tense. Tell it Through pictures You can tell a story completely through photographs or well-drawn illustrations. Pictures carefully planned and selected to form an interesting story have a very great communication value. When there are pictures, the reader's eyes get to them first. When there is a picture story, the reader would like to read it first. You can use a picture story for explaining a process or to give directions in carrying out an important practice. The picture story comes nearest to your actually demonstrating it in front of the reader. In agricultural or home Extension, a picture story can also be used for describing the 'then andnow' and 'with and without' type of subjects. A meaningful picture story is like an essay. IT has to be planned, with a theme or a central idea.

First, concentrate on the central idea. Develop the idea; get as many facts about it as you can :

All on paper. Then select the ideas, which are essential steps or parts of the story. Go over each to see whether it adds to the understanding of the story or whether the story can do without it.

Then set the steps in a logical order. If there is a particular point, which has to be highlighted in any step, jot it down so that you will not forget it when you actually take or select the photograph showing that step.

Then arrange for the pictures. Take at least two pictures to depict each step if you are taking the pictures yourself. This is to make sure that you would have at least one good picture for each step.

If you take only one, and something goes wrong with it, it may not always be possible to retake the picture. When photographs are finally ready, select the right captions and organize your story. Make a layout, where will each picture go? 'Fuse' the picture with the caption, so that they become one entity. For this, write the caption to the picture, and not merely about the picture. Many find this method suitable for writing the caption: Imagining the picture to be the lead of a feature article, and continuing the lead as the caption, to explain and amplify it. Carefully look andsee what is obvious in the picture, and what has to be explained. Explain only that. In the finished picture story; see that your pictures are all dominant. It is important to keep up continuity from picture to picture. One way of ensuring this is togive a short-write-up that relates to all pictures and gives an idea of the spirit of the group of pictures, right in the center. Keep the captions to the very minimum avoid all unnecessary words. Relate captions to the pictures by putting them close to the pictures. This is the best way ofdoing it. If you cannot do that indicate the order of the captions with letters or numbers or arrows. This will only be second best. A picture story becomes all the more interesting if your ideas are new and thoughts fresh. Try it. Here is an example of picture storyPrint media, Radio Journalism, T.V. Journalism.

Audience Research

The Audience Research, manned by professional researchers, is involved in studies on variousaspects of broadcasting. It also operates a system of getting ratings on Doordarshan programs each.

Health and Hygiene

Week Audience Research is also monitoring the voluntary feed back from viewers & maintains Data Banks at the national & Kendra level. Research methodology for audience researchResearch methodology plays an important role in proper conduct of a research and in deriving empirical findings thereof. Without research methodology the entire research structure is incomplete. Research methodology gives an idea to all concerned about various methodological steps likeas to have the data has been collected, what methods of sampling have been adopted, what statisticaltools have been used in deriving inferences through testing of various statistical hypothesis, in view of said objectives and framed hypothesis, there upon in the research study. The steps to take up Audience Research are:

- Selection of the topic
- Frame the objectives
- Review of literature
- Selection of a programme/production of a programme
  - The amount of information to be disseminated
  - The amount of broadcast time available
  - The listening ability & experience of the target audience
  - The familiarity of the target audience with the subject matter (the less familiar they are, themore programmes they are likely to need)
  - The complexity of the information to be disseminated
  - The amount of time & talent available to prepare & produce the programme)
- Development of schedules (interviews/questionnaires etc)
- Development of the scales to measure the variables
- Sample selection depending on education, age, possession of TV, caste, size of holdings, socio-economic status etc. Print media, Radio Journalism, T.V. Journalism,
- Data collection
- Statistical analysis
- Conclusion

Research Studies

Utilisation Frequency Of Media Programmes By Farm Women

J. Meenambigai & Ravichandran.

Farm women are exposed to various mass media although to a varying degree. But the frequency.

With which they utilize the media sources must be varied. In this study the patterns of mass media utilization by farm women have been studied in terms of frequency and duration of use of media. The findings shows that, even though majority of the farm women were exposed to all the media programmes frequency was found to be less. Among the radio programs morning bulletin & women's programs were listened by them in regular intervals. In television, they were frequently exposed to entertainments, farm Program & rural arts program. The present study was conducted in the ten sample villages of Madurai district. The data were collected from 120 farm women of these ten villages. Personal interview technique was used to collect data through specially structured interview schedule for the study. Frequency of Television Viewing to find out the frequency of viewing farm telecast, general & entertainment programs were selected, data were collected, mean scores & percentage analysis were worked out to interpret the results.

Mangaraj Today amidst the modern scientific developments children are habituated to manmade modernfacilities, but are little exposed to state of art of environment, when environment has passed a role ofglobal menace today. There is a need to expose the school children on such issues of environment & pollution for keeping them safe, healthy, conscious & secure on the earth as children are the futurecitizens. Bearing this, an attempt was made to study the communication behaviour of the schoolchildren in Varanasi district of Uttar Pradesh.The study was undertaken in Varanasi district of Uttar Pradesh. A total of 200 respondentscomprising school children were selected in the process of stratified random sampling after Selectingthe School, villages & blocks in the same procedure.A schedule was prepared covering the different aspects & variables for the study keeping inview pretested for assessing the validity & reliability of the schedule. The data were collected onlyafter establishing proper rapport with the students & their teachers. The collected data were classified& tabulated in light of the objectives of study & analyzed.Print media, Radio Journalism, T.V.

The study shows that the rural school children are well acquainted with radio followed by T.V. & written media. As regards to environmental awareness, they chiefly depend on news papers followedby magazines. A major percentage of them get exposure to entertainment & sports programs &minimal percentage to environmental programs in radio & Television. Much emphasis is laid byGovernment organization followed by radio, T.V., written medias, course curriculum & NGO's toexpose them to environmental aspects. In this context even though the children are concerned aboutthe environment but Government should take effective initiatives, emphasis on environmental

Education, environmental aspects utmost priority in mass Medias & NGO's should concentrate moreinterest & efforts on environmental programs. A Comarative Analysis of Farm Broadcast and Farm Telecast A. Vennila Mary & J. Vasantha Kumar the communication of agricultural technologies to the farmers is a pre-requisite of modernization of agriculture. Of the various channels of communication, radio & television are the most important mass media for diffusing scientific knowledge to the masses with great promise for agricultural development. Although radio & television cater to the needs of rural mass, it is often said that these programs do not reflect commonality or co-ordination. Hence, a study was undertaken to find out the contents of farm broadcast and farm telecast with the objectives.

□□□

Chapter 5

Agricultural Journalism : Concepts and Issues

Self-help Journalism

People who have experience aren't writing. Those who write don't have the practical experience of farming. This in a nutshell is the tragedy of farm journalism in India.

We have hundreds of farm journals for namesake. All these are published by government research organizations or agricultural departments. The writers are scientists or "subject matter specialists". They write what they think is useful to farmers. Seldom do they bother to identify the information needs of the farmers. Nor do they take pains to meet farmers and document their experiences in black and white. Most of the articles that appear in such journals are run of the mill, library-borne and of no purpose for the community for which it is purportedly intended.

Not based on needs

While writing about a crop, the writer will explain the suitable temperature range, soil condition, diseases, pesticides to be sprayed etc., but never about the actual marketing prospects or the opinions of the farmers who have already grown it. Journalism requires the writer to have commitment towards his readers. Even in the so called "popular" farm magazines, which are meant for farm readers, you don't see any concern towards the farmers. They are far from being need based.

Ban technical jargon

The scientists and experts find it very difficult to write without their usual technical jargons. As one popular feature writer has put it: "from a layman's point of view, instead of saying that a particular insect's egg is 0.25 mm, it is better to say that it is of the size of the full stop put at the end of this sentence." The scientist appears to be helpless to find out easy alternatives and simple ways of expression for his technical jargon. Difficult, it may seem, but the only alternative left to the farmer is to get educated as to what these usual jargons like "ppm", "in situ", "clone", "gene pool", "sterile", "biodiversity" etc. mean.

Lack of experience

The lack of farming experiences, especially understanding pains and stage by stage challenges involved in raising a crop or maintaining a dairy farm results in exaggerated write-ups like "cocoa - the brown gold", or in statements like "by spending 3 rupees for the mushroom spawn (seeds) you can get a return of 75 rupees". Many of our farm writings project half-truths. The farmer who is attracted by the rosy presentation of half-truth, is left to explore the other half which is more than often bitter and frustrating. These sort of unbalanced, uncommitted write-ups, written with an eye on academic credit have lost credibility among farmers. That's why even if journals which cost 8 or 10 rupees by way of production cost are offered at an incredibly low (subsidized) price of 1 rupee, no government farm magazine has a circulation worth mentioning.

Communication gap

There is a big communication gap between farmers amongst themselves, between the farmer and the outer world and between the farmer and the government. If there is a revolutionary development on the other side of ones fence, say growing cabbages without chemicals, the farmer on this side doesn't even get to know it. When the price of an agricultural product reaches very high, say tomato price jumping to Rs. 25, everybody in the society around thinks that the tomato farmer has struck the goldmine. Nobody knows exactly how much he gets and what were his problems to grow it, why the price has jumped. When there is drought in a region, the government gives fertilizers in subsidy or constructs new roads to "generate employment", without caring to know where to apply the ointment to heal the wound.

The media don't gather the farmers public opinion that they need water conservation projects to solve the crisis, not new roads. The moot question is: who is to bell the cat? Naturally it has to be the farmer, if he cares for his all-round sustainable development.

A new magazine

If farm journalism is to be need based and useful, the best way is this: farmers should start expressing through their pens. Only those who wear shoes know where it pinches. We have made a humble attempt to translate this objective into implementation. The idea was this: to encourage and train farmers to write, drawing from their field experiences.

No theoretical writings. The yardstick for acceptance of an article is that how useful or informative - in that order - it is to the farmer and not who has written it. Our association, the All India Areca nut Growers Association, a 50 year old apolitical farmers' welfare organization started publishing a pro-farmer monthly in vernacular Kannada language. The publication is not a mouth-piece of the association. All the editorial board members, including me, the honorary editor, are practicing farmers. From writing articles and editing to publishing and even selling (we have nearly 200 news agents spread over 6-7 districts) farmers play the key role. An experiment which probably has no parallel in our country.

Getting farmers to write

From the beginning, we believed in the wisdom and knowledge of our farmers. There is no dearth of it. We gave this offer: "if you have something worthwhile about farming that you want to communicate with your fellow farmer, feel free to write it. Don't bother about the style. If there is content, we will try our best to rewrite and present it, with your co-operation. We wanted to see every new development or new crop or technology or plant varieties, devices, machine etc. from farmer viewpoint.

The journalism workshops are held in calm village centers. Local farmers help in the preparations and are host to one or two trainees. In this way, there is a dialogue between farmer and trainee and the trainee gets off-class training too So when embryo transfer in cattle started making news, we had a friend veterinarian with farming background to describe and analyze the technology and give an inkling of how useful it is from an ordinary farmer's viewpoint. We call this kind of journalism "self-help journalism". We had a hope that once you create a platform for farmers to write, articles would flow like water that flows on gravity. But that was not to be. We had to dig bore wells each time to get water. After decades of being taken for granted, farmers are happy to allow others to talk and write on behalf of them. It is like a patient who has fractured one leg. He needs a lot of physiotherapy to teach his legs walking before he can resume walking on his own. Within a few months of starting the magazine, we realized the futility of pinning our hopes on known staff journalists, most of them based in cities or towns, who lack a clear understanding of farming profession and its problems. Even some in rural areas couldn't offer new content. Instead of relying on such sources, we thought it better to give some interested farmers a training of journalism.

A course for writing

So far, we have conducted 3 farm and rural journalism workshops in 3 districts, in calm village centers. A lot of pre-planning has gone into this. First, we called for applications, a sample write-up on anything farmers feel strongly about. Based on the answers to our questionnaire and their write-up, the trainees are carefully selected. Local farmers participate in a series of preparations, meetings held at village centers.

This convinces them that this is something very essential for them. They are requested to host one or two trainees during the workshop. This arrangement facilitates a dialogue and exchange of farming knowledge between the host and trainee. In the 4-day workshops, the trainee gets off-class training too. Apart from dialogue type lecture sessions on "identification of subjects", "how to write", "how to improve your writings", "development journalism", "farm journalism and the environment" etc, there used to be three writing assignments, followed by evaluation and suggestions for improvement. The last assignment is one where the trainees are taken to a nearby farm. Using the knowledge and tips they received in the last three days, they have to interview and collect information pertaining to one selected subject seen in the farm. During night the trainees write their feature. Next day, it is evaluated. During the evaluation, one by one, some trainees are asked to read their article aloud. Positive criticism and specific suggestions are invited from fellow trainers. In conclusion the panel of faculties would give their advice. If 10 trainees read their work, all the 25 will get at least 40 good tips.

The effects

Out of 76 trainees divided over three workshops, at least a dozen write often, not only to our magazine, but to others as well. The subjects they have selected are such that it would otherwise unnoticed by the media. After the workshop, a trainee reconstructed the important and unprecedented story of "acid rain" that has showered in his area a few months ago. A story that would have hit the national headlines if it was timely. He was also able to smell a subject in a new technique of bittergourd cultivation a local school teacher was employing. After reporting this story, he followed this method in the next season, met with success and wrote about it.

Another trainee could identify a subject in the "lateral pipe" that was fast replacing the troublesome "emitters" in drip irrigation. The success story of a farmer who has shown that instead of a pair of bullocks you can very well plough almost similarly by a single bullock was another local trainee's scoop. In our last 6 years of experience, there were handouts of such sweet successes. Farmers are like an old box of locked up wealth of information. Unfortunately, we ourselves don't know that we have so much knowledge. Others have made us to believe that we have to only "hear" and "read" and that we aren't capable of "talking" and "writing". If we, farmers, are able to come out of this mental block, with the active co-operation of concerned scientists and an in-depth study of the world around us, we ourselves can fulfil our information needs to a good extent. All said and done, there is still a weak point in our undertaking. Though there is good appreciation from subscribers and readers, this is not reflected in our balance sheet. Writing and sharing responsibility by the beneficiaries is not up to the mark, making the magazine dependent on the sacrifice of some.

Introduction

Agriculture is becoming increasingly informationintensive.At the same time, information and communication technology provides a range of sophisticated methods for enhancing communication with farmers. But much work remains to assess the effectiveness and feasibility of various communication strategies. Although electronic media are playing an ever more important role in agricultural communication, print media will surely remain an important source of agricultural information in many parts of the world foursome years to come. It is important to assess the degree to which media such as newspapers and magazines can play an important role in fostering the diffusion ofuseful information to farmers. This paper examines the experience of an innovative farm magazine published in India. The magazine has pioneered the idea of encouraging farmers to regard themselves as sources of material and as authors of articles, and operates with the philosophy that farmers are most interested in hearing about the experiences of their counterparts rather than the formulaicr ecommendations of 'experts'. This strategy reflects an assessment of the inadequacies of conventional farmjournalism and the limited usefulness of many publications aimed at farm households. The magazine is called *Adike Pathrike* (Areca Magazine). Its primary audience is those farmers who grow the areca palm (*Areca catechu L.*), the source ofthe areca nut (or betel nut), an important cash crop grown by smallholders in some parts of southern India. The magazine's unremarkable title and seemingly narrow subject matter conceal an imaginative approach to farm journalism and a wide scope of interest. This paper introduces the background and origins of the magazine, discusses the characteristics that distinguish it from most other farm publications (in particular its insistence on farmer reporting and verification), and examines some of the implications of the magazine's experience for farmer communication and the provision of information.

Background

Background *dike Pathrike* finds the majority of its readership in the medium- to heavy-rainfall districts of southern, coastal Karnataka and northern Kerala. The area mischaracterized by low hills and lateritic soils, making it unsuitable for the intensive rice production found income neighboring districts. Although farmers in the area grow small amounts of rice and other food crops, most of their efforts are devoted to cash crops, including areca, coconut, cashew, cocoa, and black pepper. Areca is the most important of the cash crops and has been a principal feature of local cropping systems for at least a century. The areca palm can grow to height of 20 meters. It requires well-drained soils and assured irrigation during the dry season. Once established, an areca plantation may bear fruit for 30years or more. There are two major harvesting systems employed. In some areas, the nut is harvested at maturity (when it has turned red), dried, dehuskedand sold. In other areas the unripe nuts are harvested, boiled with plant additives and dried for sale. Areca issued as a mild stimulant in a wide variety of preparations. The dried nut may simply be crushed and chewed, but by far the greater part of the production from Karnataka is sold to processors who make *paan* (a mixture of crushed areca and various spices) or *gutka* (which also includes a small amount of tobacco). A particularly popular form is known as*pen macula*, manufactured on a large scale and sold in foil packets throughout India. Areca is grown in several other parts of India, andthe market has experienced periodic fluctuations. Particularly severe drop in prices in the mid-1980scaused a crisis in Karnataka. The All-India Areca Growers' Association attempted to search for solutions. The association is based in the town of Puttur,Karnataka and, despite the name, tends to represent the interests of areca growers in this part of India rather than those from distant areca-growing areas (such as Assam), who have their own organizations. Several committees were formed through the association adhere Padre, a local journalist (and areca grower), volunteered to produce a newsletter for growers, on an experimental basis. The first several issues of the four-page tabloid were sold for Rs.1 and, with the help of advertising revenue, managed to break even. The response to the newsletter was sufficiently positive that decision was taken to explore a more extensivemagazine.Thus *Adike Pathrike* was born. It is a monthly magazine, typically of 28 pages (or more, depending on advertising). It is attractively produced, with a colourphotograph on the cover, additional black-and-white photos and diagrams to illustrate news items, and written entirely in Kannada, the local language. *Adike*

Reforming Farm Journalism

The strategy of Farmer Ledjournalism

Farmer verification

The level of agricultural communication in print mediain India is fairly high, in comparison to many other countries, but still not close to an adequate level. Agriculture is a subject of exceptionally high national importance, and many daily newspapers present stories on new agricultural technology or feature weekly sections devoted to agriculture. In addition, there are many agricultural magazines produced by state universities or government research institutes. Although government publications are dominant numerically, there is a growing number of private agricultural magazines. One of the oldest and most prominent is *Annadata*, based in Andhra Pradesh and written inTelugu, which has a very high circulation and has been functioning for more than 30 years. Although there is a wide range of publications related to agriculture, they tend to share one feature. The techniques) are almost always written by scientists, professors, or industry representatives and rely on formulaic presentation. For a new crop, for instance, there is a standard outline (botanical description, required soil type, spacing, cultivation, pests and diseases, etc.). The articles almost never discuss farmers 'actual experiences. Although they usually represent scientists' sincere and honest attempts to communicate what they have learned on the experiment station (and often contain valuable information), they lack any kind of 'ground-trotting' that would make them more relevant to the farmers' own circumstances. The recommendations for input use often include quite high doses, without reference to economic returns or environmental effects; some of the articles walk a thin ethical line by recommending particular proprietary chemicals.

Newspapers and magazines thus become straightforward vehicle for delivering a scientist's recommendations, devoid of farmer experience. Scientists find a ready outlet for their articles (some of which may be recycled material from previous years)and are not challenged to think about farm-level reality or to answer questions if their recommendations fail. Scientists often send articles to publications, expecting them to be accepted without modification. Agricultural publications, for their part, are pleased to have a ready supply of material that can go out directly under the scientist's name or be quickly modified by a staff journalist. There is little opportunity for feedback to these articles and hence no possibility of establishing self-correcting mechanisms.

When *Adie Pathrike* announced its editorial policy at the time of launching the full-blown magazine, it expected to receive many communications from farmers. But it was disappointed to find that the same *Agricultural Research and Extension Network Paper No. 128* formulaic articles from established scientists kept arriving, and that farmers were reluctant to write abouttheir experiences. The editor decided that a more proactive stance was required. Initially, farmers were requested to send accounts of any experiences they thought relevant to their counterparts and the magazine would help develop them further, in consultation with the writer, for publication. This attracted some additional material, but the experience was that in mostcases significant follow-up was required. In the first five to six years of the magazine's operation, morethan 80% of farmers' contributions required significant rewriting, editing or salvaging. The effort required to elicit farmer writing led to the idea of organizing workshops to teach basic farm journalism skills (Padre, 2001). To date, six four-day workshops (*Krishikara Kaige Lekhani*, 'handing over the pen to farmers') have been held. Most of these were given in rural communities and one was held atman agricultural university campus. The participants have included 175 farmers as well as some agricultural officers, professors and other agricultural graduates. The course syllabus is now available as a book. The workshops have been quite successful, and several of the graduates have gone on to write articles for *Adie Pathrike* or other agricultural publications. Nevertheless, only a minority of the workshop graduates are active writers and *Adike Pathrike* stillfinds it challenging to produce articles and meet its deadlines. There is a core of 30 to 40 'friends' (workshop graduates) whom the magazine relies upon.If the editor hears of a potentially interesting story, one of these people may be asked to visit the farmer who has reported the innovation, to investigate andverify it, and work with the farmer to produce an article. The author is paid a small honorarium. In other cases,farmers themselves write brief communications or atleast inform *Adike Pathrike* of an interesting story. The magazine receives many letters, often accompanied by photographs, from its readers. When this happens, the editor sends a brief set of questions to the farmer anon the basis of the responses puts together an article. Any significant contribution from a farmer-author is compensated with a small honorarium. No matter how the article is produced, it is always based on first-hand farmer experience rather than the repetition of textbook recommendations. Thus the majority of the items in *Adike Pathrike* are 'exclusives' for the magazine. *The example of manure management* Areca production requires adequate plant nutrition, with nitrogen and potash requirements particularly high. Farmers have traditionally managed soil fertility with farmyard manure (most farmers keep cattle) and green manure obtained from cutting nearby forest or fallow vegetation. For the past few decades the use of chemical fertilizer has become more common, and the majority of farmers purchase some fertilizer for their areca plantations. One of *Adike Pathrike*'s contributors, a farmer withal university degree in agriculture, became concerned with the inefficiencies in traditional manure management and sought to make improvements on his farm. Most farmers apply manure to areca by periodically clearing their cattle stalls and transporting the manure in baskets to the plantation. This requires considerable labour and does not take advantage ofthe cattle urine that drains from the cattle shed. In addition, partial dependence on the use of green manure unnecessarily destroys nearby forest land. This particular farmer has a large areca plantation and sufficient resources to experiment with new methods. He constructed a lined holding tank near his cattle shed where manure and urine could be washed each day. He then arranged a pumping system which delivered this slurry to his plantation; flexible pipe allowed the slurry to be directed to individual trees. The farmer wrote an article describing his experience for the November 1996 issue of *Adike Patrice.* The article outlined the construction of the slurry management system and described the principles contributing to its efficiency. The article also discussed areca nutrient requirements and speculated on whether reliance on slurry alone, with no chemical fertiliser,would be adequate for meeting potash requirements. The article attracted attention among readers, and in 1999 another farmer wrote a short article describing his positive experience in adapting a slurry management system. Hundreds of other farmers have adapted some form of this technique. A full-blown system for pumping slurry is beyond the resources of most areca producers. But *AdikePathrike* emphasizes principles rather than recipes, and many farmers have been able to capitalize on the ideaseatured in the magazine. Most area farmers have their houses and cattle sheds on middle ridges, while the areca gardens are in valleys. Thus they are able to arrange for some type of gravity flow to direct slurry to at least part of their garden. One farmer constructed tank from stones and cement, with a wire mesh filter to keep out debris, and hoped to use his sprinkler irrigation system to pump the slurry to his areca garden. However the pressure was insufficient, so he rigged an alternative delivery system with PVC pipe with flexible hose on the end. Since the areca gardens are lower than the cattle shed, he could deliver the slurry by gravity. He found he could deliver manure to about 400 trees per month using this system, treating his twoacregarden in a phased manner. He found that his which he ascribes to the improved fertilization. An article in the December 2002 issue of *Adike Pathrike* described this experience.

Other technologies

Adike Pathrike is full of reports on farmers' experiment sand adaptations over a wide range of farm management problems. Not all of these are concerned with crop production. Something as simple as introducing a swivel link on the chain used to tether cattle has warranted brief article. Recently a farmer reported that directing water through a length of black hose on the roof of his house served as an efficient water heater.

Reforming farm journalism: The experience of Adike Patrice in India. The magazine has also contained several household tip son combating houseflies (including the trick of leaving bowl of water with detergent foam on its surface, which attracts and traps the flies).Adaptive technology development *dike Pathrike* does not provide final answers or standard recommendations on new techniques or technologies. Instead it reports farmers' experiences and encourages experimentation and development under varying conditions. The magazine strongly believes in the wisdom and experimental capacities of farmers. Even farmers who have little formal education are innovative and knowledgeable. The magazine sees its role as providing options and pointing out the prosand cons that grass roots experience has uncovered.

Water harvesting

For the past seven years, *Adike Patrice* has featureda monthly column about harvesting rainwater. Rainfall in the areca-growing area is concentrated in one season and water both for crop irrigation and household uses often in short supply, despite the fact that many households have access to wells or boreholes. The subject of rainwater harvesting is an excellent example of the primacy of principles over recipes. The column provides descriptions of farmers' experiences in saving water, and many readers have been able to adapt the principles to their own situations. One important principle is the need to look for ways tore charge wells by encouraging the diversion and capture of rainwater that normally runs off the farm. This may involve a wide range of techniques, for instance ensuring that rain runoff from the roof has chance to collect and percolate, digging infiltration pits, or taking advantage of the rainwater reception and storage capacities of termite hills.

The development of mechanical weed control : Weed control is one of the major costs of production in areca. It is usually done twice a year, by hand. One of the magazine's readers, in collaboration with mechanic living near him, built a weedier powered ban electric motor. The electric motor was not practical, however, and they tried out a petrol motor instead. This was sufficiently successful for the farmer to begin using it on his farm. The editor of *Adie Patrice* heard about this, visited the farm and had a demonstration, then wrote an article describing the innovation. Several farmers subsequently contacted the mechanic involved who was stimulated to make improvements to the original model (for instance, linking the motor to the wheels as well as to the blade). A second article byte editor described subsequent progress. The mechanic has now made and sold a number of these welders, and his contacts with farmers have led to commissions for other types of apparatus. Few farmers can afford to own a petrol-driven weedier, butane farmer rents his out, generating income for himself and lowering weeding costs for others.

Insect control

Another example of technology adaptation encouraged by the magazine involves techniques for insect control.The editor of *Adike Pathrike* read about a method for controlling fruit fly (*Daces dorsalis*) in mango in publication produced by Kerala University. He experimented with the technique and wrote an article about it, describing how basil leaves should be ground, mixed with water, and placed in a coconut shell along with a grain of the insecticide Faraday. The shell is hung in the mango tree. The basil attracts the flies, and when they alight they are killed by the small amount of insecticide in the mixture. A little more than a year later a farmer wrote an unsolicited article (for which he received an honorarium) describing his experimentation with the technique on a related pest (*Dacus cucurbit*), thisat insect control (such as trying to protect the fruits with paper or plastic bags) and explained how he had adapted the basil-based method. He gave details on the placement of the husks and the frequency with which thebasil-Furadan mixture needs to be replaced.

Farmer-to-farmer dialogue

The above examples show how farmers are able to follow up on items in *Adike Pathrike*, refine or adapt them, and report back. All the articles contain contact information (village, postal address, and, if available, telephone). There are many instances of readers directly contacting farmer-authors for more information, income cases visiting their farms. Other ways in which the magazine promotes farmer-to-farmer dialogue include seed exchange activities, a question and answer column, and a forum for debate.

Seed exchange

Many of the magazine's readers are keen experimenters, interested in new or unusual crop varieties. Their interests include new varieties produced by university research as well as traditional varieties and those discovered or developed by farmers. As *Adike Pathrike*developed, a number of letters and articles appeared about farmers' experiences with particular varieties. Common reader response was an attempt to obtain some seed to test. Initially *Adike Pathrike* served as clearing house for these requests. It would obtain seed of a described variety from a farmer, store it in the editorial office, and send out samples to interested readers who provided enough stamps to cover postage and handling. But it became too much of a burden to turn a small editorial office into a seed storage facility,so now farmers who write to describe an interesting variety that they manage on their farms also provide contact details and instructions for acquiring small amounts of the seed. An off-shoot of this is the establishment of a seed exchange group that meets once a month to describe experiences and exchange seed. The group (*Samruddhi*) is now 10 years old and has a membership of 30 to 40 farmers. It has been instrumental in the exchange of information and *Agricultural Research and Extension Network Paper No. 128*planting material of many varieties of fruit, vegetables,ornamentals and medicinal plants.

Adike Pathrike tries to visit research organisationsand agricultural universities when new varieties preannounced. The magazine has been instrumental in helping distribute seed of new tree and vegetable varieties from Tamil Nadu University, varieties of amaranth, cowpea and tomato from Kerala Agricultural University, and has helped the Horticultural Research Institute (Tamil Nadu) test a new crop with farmers. In most cases the research organizations themselves do-little to follow up their announcements of new varieties with seed distribution mechanisms, and it is only those farmers who approach the institutes who may be ableto acquire seed.

Question and answer column

A particularly popular feature of each issue is the question and answer section (entitled 'Drops Make an are published in subsequent issues. In many instances several different responses are received, illustrating aide range of experience from the readership. For example, a reader's question in January 1995 concerned how to control the gundybug (*Leptocorisa acuta*), sucking pest of crops such as beans and cowpea. The March issue published five different responses. A retired university entomologist wrote to outline the life cycle of the insect and to suggest how to find and destroy its eggs by hand. Another scientist wrote to explain that in the morning hours the gundybug is inactive and can be found and destroyed. A farmer described his positive experience with spraying a mixture of onion juice and water every three or four days. Another farmer explained how an intercrop of marigolds in thevegetable garden limited damage from the pest. Perhaps the most ingenious contribution ICTs are basically information handling tools - a varied set of goods, application and services that n was from a farmer who described how red ants can be encouraged toattack the gundybug. If a red ant colony can be located near the garden it is possible to provide the ants witha 'highway' (a length of rope) from their home to the garden and a potential gundybug meal.

Other examples of dialogue

The exchanges published in *Adike Pathrike* may not lead to definitive answers, but they serve to alert farmers to possibilities warranting further examination. Vanilla has become an important cash crop in the area in recent years, and farmers have much to learn about its management. For instance, there is controversy over the length of cutting that is appropriate for propagation and how to manage it. A recent article reported farmer's experiment on the direct planting of vanilla cuttings versus a potting stage. The article found the latter method superior, but other farmers have disagreed.

A counterbalance to promotional campaigns

The areca-growing area is particularly sensitive to changes in markets for cash crops, and farmers areaways on the lookout for new opportunities. There are frequent instances in which either the government and/or private industry launch campaigns to promote new crops or enterprises. The campaigns often feature some type of introductory subsidy or other incentive to encourage farmers to take up a new enterprise. Not surprisingly, some of these eventually prove to be genuine opportunities and others are dismal failures. The conventional agricultural newspapers and journals generally report only the positive side of these stories, acting as a conduit for government or industry press releases and other publicity. Exaggerated production figures may be cited and risks from insects, diseases or inadequate markets may not be explored. There are few instances in which the press looks critically at the new opportunities or reports failures (save in thecae of a spectacular scandal that might be judged newsworthy).

Oil palm

Adike Pathrike takes a much more even-handed viewof such campaigns, and in doing so stimulates farmers to look carefully at new opportunities and compare experiences. For instance, in the mid-1990s the government began an ambitious campaign (in conjunction with private industry) to promote the cultivation of oil palm. India's demands for cooking oil exceeded its supply and the government was anxiousto develop alternatives to the traditional oilseed crops. The oil palm initiative was introduced with a great fanfare and participating farmers were provided with seedlings and subsidised inputs, at least in the initial stages. A number of very positive articles appeared in the local press describing the farmers' satisfaction with the new crop (even though in most cases it had yet to bear fruit). However, *Adike Patrice* began to receive reports of farmer dissatisfaction with the crop; the initial supports and subsidies had evaporated and participating farmers were having a hard time making ends meet as they waited for the palms to come to maturity. None of this dissatisfaction was reported in the press, so *Adike Patrice* decided to follow up one of the 'successful' farmers who had been featured in a newspaper report. They were surprised to find he had removed most of his oil palm and was deeply in debt. The magazine published an article on this experience(and subsequently a rejoinder letter from one of the companies promoting oil palm). The oil palm enterprise in Karnataka is not a complete failure. Many farmers gave up on it, but a number went on to produce profitable crop. But the case illustrates the fact that farmers generally have little help from the conventional press in looking objectively at new opportunities.

Tissue culture

A second example involves the growing availability of tissue culture technology in India. Although the techniques of tissue culture have been understood for some time, it is only in the past few years that the technology has become potentially accessible to the average farmer. There are now more than a dozen *Reforming farm journalism: The experience of Adike Pathrike in India* commercial laboratories in Bangalore (the capital of Karnataka), offering tissue culture plantlets of species such as mango and banana. Tissue culture is wonderful example of 'high science' in agriculture ands such receives many glowing reports in the press. However, there are few places that farmers can turn to understand the potential and the limitations of the technology. Following its strategy of providing farmers with the opportunity to describe their own experiences, *dike Patrice* published its first article on tissue culture in 1998, in which a farmer described his positive experience with both tissue culture mango and banana. Another contributor had heard of several failures in tissue culture banana, and went to visit four farmers who had negative experiences. An article appeared in1999 (accompanied by a photo of one of the disappointed farmers beside an under-producing tissue gone wrong (including the possibility that the laboratories had supplied inadequate planting material).The exact nature of the failures has not been completely resolved, although one of the farmers involved has taken the tissue culture laboratory to court. More recently, in 2002, another contributor provided experiment with tissue culture banana. Of equal importance, this article provided tips to farmers who are considering going into banana as a cash crop(including choice of variety and time of plantation establishment) and gave guidelines on the scale of planting that would justify investment in tissue culture allows farmers access to some 'real-life' experiences to contrast with the uniformly rosy reports appearingin the conventional press.

A loan scandal

Although *Adike Pathrike* normally looks at only the technological and economic merits of new enterprises, there also have been a few cases of investigative reporting uncovering outright fraud. One example involved local cocoa farmers being encouraged to sign up for loans through a cocoa-buying company. The company was taking advantage of the availability of farm loans and had collected the signatures of thousands of farmers. The loans were either not delivered to the farmers or arrived late, and it emerged *dike Patrice* published an article on the situation and the All-India Areca Growers' Association was ableto pursue the case with the government (as many obits members are also cocoa growers). This is one ofthe few examples of muckraking by the magazine, and the successful pursuit of the case owed more to the willingness of the growers' association to take it on than to pressure from the magazine's readership.

Conclusion

Although the story of *Adike Pathrike* is confined tone small corner of India, the experience of thisfarmerledmagazine has a number of implications for farm journalism in developing countries. This concluding section reviews the impact of *Adike Pathrike*, examines some of the factors that condition its success, reviews some of the challenges the magazine faces in its own environment, and speculates on the future of print media for agricultural development.

Impact

There is no doubt that *Adike Pathrike* is a success. It has continued publishing for 15 years and stands on its own feet financially. It has attracted a loyal readership, and a recent reader survey indicated thatmany farmers save and bind past issues rather than discard them. It is one of the few examples of an agricultural publication based on farmers' experiences and encouraging farmer-to-farmer contact. A number of the innovations first described in the pages of the magazine are now widely adopted among farmers inthe region.Its unique approach has been quite widely influential. There are now other publications trying toemulate the farmer-led reporting of *Adike Pathrike,*and a number of the graduates of the writers 'workshops are able to publish articles in other newspapers and magazines. A recent promotional event by the government Spices Board included not onlythe usual presentations but also invited farmers with experience of growing the crops a chance to present and discuss their experiences. In another case, *AdikePathrike* (in collaboration with the All-India ArecaGrowers' Association) was asked to organise a meeting on the subject of micro-irrigation in which not only outside 'experts' presented their findings but experienced farmers also were invited to discuss the pros and cons of various techniques. The magazine recorded the proceedings of the meeting and produced summary publication.

Conditions for success

The *Adike Pathrike* model provides a number of fresh ideas about the content of farm publications and the importance of involving farmers in the provision and review of information. However, it is important derecognize some of the characteristics of the environment in which *Adike Pathrike* operates that may limit the direct transfer of this model to other situations. In the first place, India is one of the few developing countries where it is relatively easy to establish a new publication of this type. The country has a rich and varied publishing tradition, and there are many newspapers and magazines, both national and local, published in a variety of languages. Even the smallest town has a news stand and often a bookseller. The relative cost of a newspaper (e.g., as a proportion of the daily agricultural wage) is one of the lowest in the developing world.

Adike Pathrike is edited and composed in a small town (Puttur) and then printed by an experienced publisher in the city of Mangalore, about 50km away. The postal system in India insufficiently efficient to ensure the delivery of thousands *Agricultural Research and Extension Network*of copies of the magazine to subscribers each month,and a network of booksellers provides an additional outlet for sales. Another conditioning factor is the type of farmer who reads *Adike Pathrike.* Although few areca growers could be described as large farmers (farm sizerangesfrom half a hectare to two or three hectares, with avery few holdings approaching 10 hectares), the majority of the readership comprises farmers who have sufficiently profitable and diversified holdings to support viable farm households. The readers obviously have at least a minimum level of literacy, while many have completed secondary school and a considerable number have university degrees.

These are also farmers who have access to a wide and growing range of communication media; many readers own television sets, for example. This is not to say that the magazine caters to the needs of the well-to-do, and many of the ideas and innovations discussed in the magazine are taken up and applied by farmers who are not regular readers but have seen or heard about the innovations from neighbors. Nevertheless, the success of apublication like *Adike Pathrike* almost certainly depends on the existence of a core of fairly well educated and economically viable farmers.Despite the loyal support of its readership, it is doubtful if *Adike Pathrike* could survive in the absence of advertising revenue, without which the price of the magazine would have to be doubled. Although areca(and some of the other cash crops grown in the area)require relatively small amounts of external inputs (andin any case the magazine refuses to accept advertising from pesticide firms), there are many other products used by farm households that can be advertised. Areca farmers are purchasers of pumps, sprayers and motors; they are interested in a wide range of small-scale machinery and equipment, such as that used for microirrigation;and they have an interest in products such as organic fertilizer, seed and areca seedlings. The farming system is thus sufficiently well developed that wide range of enterprises find it worthwhile to advertise in the magazine. Finally, the farming system itself is supportive of opportunities.

The area is largely one of cash crop production and farmers' livelihoods are linked to markets. Farmers are interested in hearing about ways of improving their efficiency, learning about trends in markets, and obtaining insights into new enterprises. Beyond these strictly economic conditions, the farmers in the magazine's catchment area grow a wide range of fruits and vegetables for home consumption and income cases for market, and many take an interest in experimenting with new varieties or reviving old ones. A significant proportion of the readers enjoys tinkering with equipment or machinery and is on the lookout for new ideas. Thus *Adike Patrice* exists in an environment of fairly remunerative cash cropping, on-farm crop diversity, and a tradition of experimentation.

Limitations

Although *Adike Pathrike* is certainly a success, there are several factors limiting its further expansion. These include the status of the areca market itself, socialorganisation in the countryside, the struggle to attract capable authors and emerging competition. *dike Patrice* was born during a time of crisis in the industry in the mid-1980s. By the early 1990s the trend was reversed, areca prices were good, and areca area was expanding. The initial crisis triggered a search for solutions and encouraged people tread the early version of the newsletter. As the market recovered, an increase in farm incomes encouraged more farmers' to take an interest in the new magazine. The magazine itself contributed little to the upturn in the areca market, and continues to be a valuable source of information on alternatives to areca, but its fate is in some sense linked to the incomes of local farmers. In the past few years areca prices have dipped sharply again and the consequent drop in farm incomes is certainly related to the magazine's relatively flat sales performance inthe past few years. The settlement pattern of farming communities in the areca-growing area may also impose some limitations on the success of the magazine, or other agricultural media. The nature of the farming system (carefully managed plantations combined with animal rearing) means that farm households are dispersed and there are fewer opportunities for communication among farmers than in nucleated communities. Dispersed settlement is of course typical of many farming systems, and it limits the spread of information. A magazine like *Adike Pathrike* serves an important role among its readers, and there is evidence that the magazine silent and shared with others, but with the dispersed settlement pattern there are fewer opportunities for this than elsewhere. In addition, the dispersed settlement may contribute to a relative lack of politicalorganisation among areca farmers. The producers 'association is a modest operation and tends to mobiliseonly in times of exceptional need. There are currently few other modalities through which farmers might defend their interests or lobby for improvement. Recognition of the significant innovation of supporting farmer-journalists must be tempered by an acknowledgement of the constant struggle of *AdikePathrike* to develop material. The magazine has come long way in establishing a reputation and stimulating correspondence from farmers. But it is rare to receivecomplete, unsolicited article in the post. Most of themagazine's articles are generated through the editor and then developed by the group of farmer-journalists (who receive an honorarium for each article). The editor must be constantly on the alert for new material andcontacts.Developing this small cadre of collaborators is significant achievement, and one which others might well attempt to emulate, but it does not overturn the observation that it is very difficult to get farmers to write about their own experiences.

Reforming farm journalism: The experience of Adike Pathrike in India

Finally, there is emerging competition for *AdikePathrike.* Two other agricultural magazines have recently started up in Karnataka. One is sponsored ban NGO and aims at farmers and farming systems inthe northern part of the state; it attempts to follow some of the farmer-driven philosophy of *AdikePathrike.* A second magazine, run by a family of agriculturalists from Mangalore and following *AdikePathrike*'s pro-farmer approach, has started publishing. Another farm magazine in Kannada, from the publishers of *Anna data,* is also in the pipeline. A radio station in Mangalore has begun to broadcast a half-hour farmprogramme, using some of the philosophy (and farmer reporters)*Adike Pathrike.* For farmers with more resources, additional options are available. The farmmagazine *Annadata* (published in Telugu in Andhra Pradesh) has branched into television and now produces a half-hour weekly agricultural programmebroadcast on the local Kannada-language channel. For those with access to the internet, anew website representing the Centre for Alternative Agricultural Media (CAAM) (www.farmedia.org) has recently beenestablished. The website is in English and attempts tobe relevant to all Indian farming;

it adopts a number of the principles of *Adike Pathrike* (promotion ofeco-friendly technology, emphasis on principles rather than recipes, reliance on farmer-based experience).Developing effective print media to serve farmers' interests

The experience of *Adike Pathrike* should cause reflection about the role and purpose of the printin furthering farmers' interests and agricultural development. It points to glaring inefficiencies in technical communication, underlines the interactive nature of farm communication, and poses questions regarding the information requirements and priorities of farm households. *dike Pathrike* provides, in its own quiet way, astunning condemnation of most farm journalism and indeed of much communication from agricultural research. It is important to remember that this examples found in India, a country with an exceptionally high commitment to agricultural research and an extensive network of agricultural universities and research institutes staffed by well-trained and committed personnel. Nevertheless, the record of communication between the experiment station and farm leaves much to be desired. Part of the success of *Adike Patrice* is based on addressing farmers' frustration in receiving only formulaic advice, recommendations devoid of anyone-farm testing, and a non-existent seed supply forthe new research varieties reported in the newspapers. The case is even more serious when one realises that the *Adike Pathrike* readership must be described as'middle class'. The magazine's clientele is not composed of illiterate farmers struggling for subsistenmarginal land, but rather well-educated, articulate, successful cash-crop growers. If they are frustrated withthe relevance and quality of communication from public research, what must it be like for the vast majority of India's farmers who are less well connected andconfident?This discussion should not be construed as challenging the need for producing standard recommendations and technical literature to report research results. But the experience of *Adike Pathrike*underlines the great gap between this type of information and farm-level decision-making. Farm journalism has, by and large, seen its role as simply transmitting, or at best translating, these formal publications to farmers. The effectiveness of such information transmission is much in doubt. Farming isan adaptive, experimental, idiosyncratic endeavour, and *Adike Patrice* has demonstrated the need for farmers to test innovations and compare experiences. Of course, farmers around the world do this, through discussions in the village or observations of neighbours' fields. Farm journalism has the opportunity to promote this type of farmerto-farmer discussion across a much broader area, but has so far not met the challenge. Part of the failure to meet this challenge can be blamed on unimaginative or poorly motivated journalism, but there are other factors as well.

Farmers,it must be acknowledged, are often not the most communicative people. *Adike Pathrike* struggles to elicit materials and depends heavily on a small group of farmer-journalists. The system works in this case because of a spirit of communication and shared interest in the areca-growing area. Such a spirit is forelocking in farming areas and limits the possibilities for developing innovative farm journalism. In addition, itis important to ask what role agricultural research should adopt. The top-down provision of recommendations is clearly inadequate, but there should be opportunities for research to learn from the discussions that take place in the pages of a magazine such as *Adike Pathrike* and help move towards syntheses (based on principles rather than recipes) that farm journalism by itself would have difficultymanaging. Finally, the experience of *Adike Pathrike* causes reflection on the priorities of farm journalism and the type of media needed to strengthen farming communities. The emphasis is usually on technical recommendations, packages of practices and new technology. While not challenging the role of such prescriptions, *Adike Pathrike* has demonstrated significant demand for the opportunity to share experience, to stimulate experimentation, and to take broader view of the farm household. The magazine has featured important contributions on new crop opportunities and the latest technology. But surely an important part of its success is that these innovations are discussed alongside more mundane issues (such as controlling houseflies or providing hot water) that contribute in an important way to the quality of rural life and to the incentives for pursuing technologicalchange. *Agricultural Research and Extension Network Paper No. 128*

Farm Journalism

Farm journalism today deals with the dead. It is no longer about the living. Ask journalist Palgummi Sainath, he'll feed you with inputs on farmer suicides. Ask economist Jagatai Ghosh, she'll spice up your palate with details on the emerging Indian food crisis. Ask the farmers, they'll grind in their desire to quit farming.

It is, therefore, not surprising to see agricultural scientist and food policy expert, Professor M.S. Swaminathan, making a feverish appeal to the media to help farmers feed 1.15 billion Indians in addition to a billion farm animals. (The media and the farm sector, M.S. Swaminathan, The Hindu, November 11, 2009) says that Indian farmers collectively need to double their annual cereal production from the current 220 million tonnes in order to feed over 1.8 billion population plus farm animals by 2050. Is it possible? It has to be made possible, says Swaminathan, because India promised, as part of the UN Millennium Development Goals, to halve the number of the nation's hungry by 2015.

Forty years ago, when global media and food experts gave up hope on India's ability to feed itself, avoid starvation and hunger deaths, the nation miraculously avoided the mass graveyard by sowing 18,000 tonnes of high-yielding Mexican wheat and rice seeds. Thus began the first Green Revolution in India. The imported semi-dwarf varieties of wheat and rice seeds greatly enhanced the yield and incomes. The nation's food grain buffer stocks swelled and everybody (almost!) had food to eat. Charismatic (former) Prime Minister Indira Gandhi's biggest political gamble (garibi hatao!) of the 1960s paid off. As a result, Ms. Gandhi's Administration pumped in more funds for farm research and food scientists led by Swaminathan, who were the key architects of the Green Revolution, developed newer varieties of high-yielding high-breed seeds. There was, however, one glitch. Enormous quantities of chemical fertilisers were needed to aid the new seeds to give high yields.

This is where the media, especially All India Radio, played a major role in the success of the Green Revolution by bridging the gulf between government policies, extension services, new technology and eager farmers. More than the English-language media, the language media had done a yeoman service to the Indian farm sector.

Despite its successes, Green Revolution — the term was coined by Dr. William Gaud of the Department of Agriculture, United States, in 1968 — triggered angry debates. Extensive studies had shown that the excessive use of chemical fertilisers over long periods killed the nutrients in the soil. A dead soil cannot grow food grains. Farmers in states like Punjab and Andhra Pradesh, who were already suffering due to erratic monsoon, realised this the hard way. Dead soil was one of the several factors that ultimately led to the second round of food crisis that is now staring in the faces of Indian farmers and policy makers alike. Hence, the renewed interest in the Second Green

Revolution. Prime Minister Manmohan Singh expressed his desire for the round two of the GR and his food policy adviser Swaminathan is eager to orchestra the media support for revolutionising Indian agriculture once more.

However, it does not seem necessary to demonstrate the importance and timeliness of farm journalism as a subject of intellectual curiosity and public concern. The Indian media, after four decades of trying with increasing success to isolate agriculture, has begun to shift its focus to food crisis and food security. One of the key drivers prodding media to tilt towards agriculture is the global interest in the commodity market. It is another issue that trading in food grains is pushing farmers into a new world of speculation, debt and uncertainty. Media today is not bothered about the actual process of farming or the wellbeing of farmers. Its interests lie in the financial implications.

Senior journalist S. Viswanathan rightly states that the "... character and composition of the news media have been ... transformed. Growing urbanization and metropolitanisation, and the hyper-commercialization of the media in the context of the neo-liberalization policies of the government are the key drivers of this big change. Relevant rural coverage, except in a small minority of newspapers, has become minimal. This does not of course mean that the rare species of committed progressive journalists is moving towards extinction." (From green to evergreen revolution and roles of the news media, S. Viswanathan, Readers' Editor, The Hindu, November 16, 2009) We hope so.

This journalism is about growing

Shree Padre delves into the details of a unique, successful experiment of self-help farm journalism.

Those who grow never write; those who write don't grow. This, in a nutshell, is the root problem of farm journalism. India has a good number of farm magazines, supplied at highly subsidized rates through subscription. Their aim: betterment of farming community. In numbers, they look impressive. But what about quality and purpose?

This article is about the details of a unique, successful experiment of self-help farm journalism, that is, farmers penning for their fellowmen. But before that it is essential to examine the deficiencies and drawbacks of the present kind of farm journalism.

Farm journalism is being fed mainly through two ways. *One:* through the magazines. Nine out of ten such magazines are either published by our agriculture universities or by the agriculture research institutes run by Indian Council of Agriculture Research. *Two:* through farm pages of mainstream dailies and periodicals. The commonality in both: the writers. Either they are scientists or professors of these institutions or so called agricultural experts. In one way or other all these writers are government controlled. These gentlemen write what they believe is useful for the farming community. But, in reality, for the target groups, it's nothing more than an academic output, which remains far off from their information needs.

Library-borne write-ups

Most of these write-ups are library-borne. Field contact is conspicuously lacking. That's why, a feature titled, *Profitable crop:* Turmeric is almost a carbon copy of what has appeared a couple of years ago; except the long list of 4-5 writers attached to it. Old wine in old bottle. In the fields, cultivation methods differ from plot to plot; they don't remain stagnant for years. But even after a decade, our experts present the same old turmeric article as if it is new. Writers stick to a hackneyed formula. Take any crop, there are fixed sub-headings. Botanical description of the plant, required soil type, spacing, cultivation method, irrigation, manuring, diseases and pests, plant protection and economics. There is so much about the crop, but unfortunately, so little about the farmer.

What does the farmer have to say about this crop? What are its merits and demerits according to her/him? How does the method compare with older crops? Is there dependency for seed/planting material? Is this crop labor intensive? How easy or difficult the marketing is? Does it need profuse irrigation? How is the incidence of pest attack and diseases? Is pest/disease management difficult? How are the prospects of the crop assessed? Aren't these unanswered questions important from readers' point of view? Does your statistics of 'income per hectare' picked from research plots have any relevance for him?

Most of the articles recommend a heavy dose of chemical fertilizers and pesticides. Much against the ethics, many a times, brand names are also mentioned. This leaves a reader wondering whether the writer is doing a public relations exercise for some chemical company. Even if there are some proven non-chemical methods adopted by farmers, they are acknowledged, let alone being included.

Presentation of half-truths

Exaggeration and passing half-truths as reality is another serious problem. A farm journalist should have more responsibility and self-restraint before going into black and white. Thousands of farmers read an agricultural article, put lots of expenditure and pains to implement the recommended idea. The results are often not immediate, may be in a year, or even half a decade after. Whenever an article on a new crop or latest technology appears, it seldom makes any mention of the demerits. Only after years, when next crop is released or when improved version of this technology comes, the experts reveal the drawbacks of earlier crop/technology to emphasize that the newer is better. Farm writing of this sort misleads readers.

Furthermore, scientists are unable to communicate simply and effectively. But to be fair, during the course of their education, with the exception of extension staff, scientists are not trained for this. I am always of the opinion that the curriculum for agricultural scientists should include training to equip them with better and simple communication methods, and in local languages. Interaction with target groups of farmers then becomes less reliant on extension staff.

Adike Pathrike's efforts

Being let down by the farm articles many a time, we, an apolitical farmers organization, All India Areca Growers Association®, started a publication of our own called *Adike Pathrike,* over 12 years ago. Adike Patrice was initiated as a monthly in Kannada mainly catering to areca growers. The uniqueness of the publication is that all major disciplines of a magazine production such as editing, publishing and even distribution (news agency) is mainly shouldered by practicing farmers, in addition to writing articles. This is an adventure which perhaps have no parallels in the country.

Adike Pathrike addresses the areca growers information needs. One advantage we have is that being farmers ourselves, we are in fairly better position to precisely list out our information needs and to divide it according priority. Our major challenge lies in identifying the writers who can deliver right kind of information, and then, to get it written by them.

The subjects we cover are of a very wide range. It's not about arecanut cultivation alone. Animal husbandry, dog rearing, poultry, research news, other crops like coconut, cocoa, banana, pepper, spices, vegetables, fruit varieties, farmer innovation, experiments, indigenous knowledge, herbal medicine, labor saving devices, farmers organizations, marketing achievements, organic farming, non-chemical pesticides, manure preparation, success stories, failure stories, post harvest technologies, driers, environment, rain harvesting, soil & water conservation (S&WC), forgotten agriculture practices – the list is endless.

A meaningful platform

When we started, there was no such open platform for farmers. We thought that articles would pour in because we sincerely intended to highlight farmers' own time-tested experiences, That was not to be. What poured in were the same formula articles, which had a big name tag and very poor content. It didn't take us long to realize that to provide need-based information, you can't rely on write-ups that come on their own by mail. You have to slog to achieve quality.

We started appealing to farmers to send in any write-up, which they thought has a new and definite message for their counterparts. If need be, we'll rewrite it, we announced. This melted the ice, but not to the desired extent. One thing we were convinced was that if there is content in the output sent by somebody, style can be given. It didn't' matter if the situation required 2-3 rounds of correspondence or a few telephone calls. In the initial 5-6 years, more than 80 percent of farmers' inputs required subbing, rewriting or salvaging depending on the contents. But then, it is worth all the trouble taken.

Another realization that came is this: Unless the writer has a thorough knowledge of farming, or still better, experience in farming, it is very difficult to do justice. This is because a farm article calls for lot of practical ideas. It requires judgment, putting right questions to your source at right time, eliciting the information etc. Somebody has wisely said that only those who wear shoes can pinpoint where exactly it pinches!

Training farmers in journalism

Instead of pinning our hopes on staff journalists who didn't have time or inclination or both to pen down a worthy article for us, we wondered, why not impart a working knowledge of journalism to enthusiastic farmers? This is the objective with which we started our *Krishikara Kaige Lekhani* (meaning handing over pen to farmers, in Kannada) workshops. We gave wide publicity. Trainees were selected from among the applicants after screening.

Thus far we have conducted six four-day workshops in some selected rural hamlets covering 5 districts. One was organised in GKVK campus of University of Agriculture Sciences, Bangalore. Workshops covered lectures, exercises and a field assignment to identify issues and to elicit right kind of information from fellow farmers. Apart from 175 farmers, a sprinkling of agriculture officers from banks, professors from agricultural universities, a good number of agriculture graduates have been trained this way. The orientation is not only to write to our magazine, but to the mainstream media as well. Several writers have straddled the rungs of journalism and are today writing for even bigger publications.

The syllabus of the KKL workshops is brought out in a book form under the title Krishikara Kaige Lekhani. Today this has turned a reference guide to all the farm journalism workshops.

A view of agricultural journalism workshop

Instead of 'producing' an issue from the voluntarily sent articles, we started identifying and commissioning farmers and others who had the stuff. If they were entirely new to writing, we would send a threadbare questionnaire. Answering all the questions almost provided the info you were looking for. If the writer was our trainee, some guidelines here and there would suffice. The lions share of our contents are exclusive and commissioned. If need be, we contact the commissioned farmer again and again by post or mail. On special situations either us or our trainees make a spot visit to help prepare a more reliable and objective article.

Identifying the information needs, searching for writer, contacting, encouraging him, follow-up - this is always an ongoing task. If you make a contact today, the article in its final shape might get ready after 6 months or one year. Our editor's job is not restricted to 'editing' only. He is more of a visualizer, an organizer, a guide and sometimes even a 'pesterer'!

Whenever we publish about a new crop, we juxtapose the introductory article with an opinion oriented one from an experienced farmer who has grown it for at least some years. The contact address and telephone number is also published. The idea is to provide the scope for an interested farmer reader to have a first-hand study and to come to his own conclusion.

Similarly in farmers' meets and seminars where we participate as co-organizers, we suggest inviting experienced farmers also as guests along with experts. There should be a mutually respecting relationship between the science and experience. It's high time to change the mental block that 'all experts are here to say and farmers to listen'.

Different information dynamic

The farmer is an unavoidable source of information because he is the real performer. Moreover, the question of credits, policy restrictions, institutional permissions and reputation and other issues does not come in his way of revealing the truth. The farmer calls a spade a spade. Even if some farmer tries to exaggerate or to plant stories with a vested interest, other farmers criticize this in the very next issue, though they might choose to ignore such misinformed writing by a scientist.

If a new crop grown by a farmer fails, s/he never conceals it. Most farmers are transparent in their views. If a newly wed bride asks with a mother of two as to how the delivery pain is like - the latter won't lie. Seeking an experienced farmer's views can be likened to this. For other farmer readers, a farmers direct views are always easy and helpful to arrive at their own opinion.

The approach of our agriculture extensions in Institutions/ Universities does not take note of the wisdom, thinking capacity and intelligence of the receiver. Moreover, it's always one sided. A farmer's innovation or invention, however successful it is, not picked up by extension network for further spreading. Instead, it prefers to impose an unsuccessful one from a government research centre.

Self-help extension

Over a decade of observation, we found that for many of our farmers problems, their own counterparts have the right answer. Because of our poor farmer-to-farmer communication, this potential is not realized by the higher-ups and extension pundits. We, for example, have a monthly column called "Readers' Questions" in which farmers put many questions they are otherwise unable to answer. Some questions could even be silly sounding, but may be very important for their profession. Questions might range from how to control crab menace, to why coriander available in market doesn't germinate. Interestingly, other farmers come out with answers most of which are very useful. This column of Adike Pathrike is very popular.

An agricultural publication like many other profession-based journals has three possible expectations from its consumer. Buying, reading and most important - implementing. We get a feedback of considerable amount of implementation of practical ideas carried in our magazine. By way of diffusion, there are instances where it has extended to non-readers too.

Electronic fence to prevent cattle and wild boar, control of crabs, fruit-flies, rhinoceros beetles, vermicomposting on home and commercial scale, growing paddy on your threshing yard under rain fed conditions, 'ringing' method to induce yielding in citrus plants, pruning of mango trees for yielding, growing vegetables by double-digging method, top-working in jack & mango, emphasis on growing multiuse variety of forest spices like Kokum (Garcinia indica), Uppage (G.gummigutta), protecting areca gardens from northern sun-rays, recycling dried areca leaves etc in garden itself, fodder production for dairying, liquid slurry usage instead of buying manure from outside, sand bag check-dams, rain harvesting, and so on — the list of widely adopted ideas is still big.

Creating awareness

By painstakingly collecting and disseminating information — even through slide-shows and demonstrations — we are able to build up considerable awareness in our circulation area of 8 districts - where water recharging was an unknown word earlier. For the past four years, we have been running a feature series in the subject. These are mostly success stories collected from different parts of the country. Taking clue from the series, quite a few people have implemented rain harvesting in their own way. Success stories have come from our readers themselves.

Adike Pathrike published a feature on easier method of paddy cultivation based on a farmer's experience. Now this is practiced by many farmers in the region

The interest is growing. One outstanding success story has come from Soraba. Mr.Lakshminarayana, a Junior Engineer in State Electricity Department had had a 30 ft well dug. It was almost a failure

- with only one feet of water. He then started harvesting rain. Spending 1,200 Rupees, he had 4-5 long contour trenches dug in his 2-acre plot. That changed the whole scenario. Last year, 3 seasons later, when Lakshminarayana wrote the feature on our request, his well had a pretty 16 ft of water in summer end.

Adike Pathrike has a column called seed distribution. Agricultural Universities and Research Stations release new plant varieties. But the interest with which they send the press releases for publication doesn't sustain; in the meantime, interested farmers approach them for some sample seeds. We tried to bridge this gap in a small way. We collecting information on newly released seeds from scientific institutions as well as on threatened varieties from farmers. A publication on the variety follows in the magazine. Seeds are then distributed by mail to those who send self-addressed envelopes and directly to farmers who visit our office for the purpose.

Widening the canvas

Later, we started two important columns: Plant Friendship and Farm Experience Exchange Tour. Like a pen friends column, readers with plant varieties and those interested in having some specific plants send their entries to the column. After publication, there will be mail and personal contacts, developing into friendship, courtesy a plant.

Imagine that a farmer wants to make a study about a particular crop or technology. It is no use directly visiting the place, staying in nearby lodge and then making inquires. Instead, if s/he directly visits a farmer of that area with prior consent, the visit would be more effective. This way farm information is exchanged between those interested to 'talk about it' and those who want to 'know further'.

Five years ago, we floated a unique plant friendship organization at Puttur called 'Samruddhi'. The objective is to exchange information and planting material between hobbyists of vegetable, fruit, ornamental and medicinal plant growers. About 40-50 farmers from the neighboring five taluks meet regularly in a month. Exchange takes place in the meeting and also personal contact between members. Newer varieties of plants, seeds etc are 'imported' from other parts of the country and are distributed. Study tours are frequently conducted. Variety consciousness has risen a great deal. For plant-lovers around, Samruddhi meeting is a much-awaited occasion now.

The primary limitation of farmer-to-farmer communication is that somebody has to zealously encourage and follow-up assignments. Many farmers have hesitation and an inferiority complex. They have to be reassured. Most farmers are happy if someone volunteers to do the writing work for them. Payment or things like cover page exposure are not much of an attraction for farmer writers; nor do they understand the pressure of deadlines. There is also a tendency to generalize and to jump to conclusions. But all said and done, practising farmers have a pandora's box of information - for those who are able to open and utilize it. Still, even though getting farmers to write for their counterparts is difficult, it is not impossible.

Over a decade of standing, Adike Pathrike has been successful in making a definite footprint in purposeful farm journalism. Many of the district newspapers reproduce articles from our magazine, mentioning the source. Mainstream press is now accommodating more agriculture articles. On invitation from Krishiloka, a Kannada farm monthly of 25 years, Ksheerasagara, a Kannada monthly on dairy farming by Karnataka Milk Federation (KMF) and the Kerala Horticulture Development Corporation (KHDC) which publishes a newsletter in Malayalam, the Adike Pathrike team has planned and conducted journalism workshops. Honeybee (English - Gujarat) and Iliea (English - Netherlands), Malayala Manorama (Malayalam - Kerala) and many Kannada periodicals have published features hailing the idea of farmer-to-farmer communication. An opinion poll conducted by us found this: most subscribers of Adike Pathrike don't put their old issues to 'raddi'; they are book bound and preserved for future reference.

Reliability, it seems, is our greatest achievement

The Concept of Innovation Journalism

"Innovation Journalism" is the intuitive name for journalism covering innovation. It is valid as a concept, but there is no such journalistic discipline or community today. Therefore, a programme has been designed to develop the concept and test the possibilities for Innovation Journalism as a journalistic discipline, a new reporting "beat".

A Google-search on "innovation journalism" generates zero relevant hits apart from the present program.[i] In comparison, "business journalism" generates 18400 google-hits and "technology journalism"

5120 hits. However, when suggested in discussions, some editors are willing to refer to their publications as "innovation journalism", and I have found prominent journalists who are enthusiastic about calling themselves "innovation journalists". The purpose of the international program is to give some of these editors and journalists an opportunity to develop their innovation journalism skills, and to form a community. In a year we might know if "Innovation Journalism" may be a journalistic discipline - a new reporting beat.

VINNOVA is managing the program in co-operation with the Swedish Foundation for Strategic Research[ii], the US Council on Competitiveness[iii], and Profnet[iv].

The concept of Innovation Journalism

Many important companies develop and sell innovation-based products today. In order to discuss the future of such a company, it is very important to assess their innovation. How strong is their R&D? How strong is their innovation management? Will they be able to earn money on their innovations? These questions calls for investigation and analysis of the links between technology trends, R&D policies, immaterial property rights, investments, technical standards, industrial production processes, marketing of new technologies, business models, politics, and more.

If reporters are to offer a well-founded opinion about whether or not a society or company is heading towards growth, they need a thorough understanding of both technology and business matters. A piece of good technology with bad business management is a guaranteed failure. A piece of bad technology with talented business management may create false expectations, where many investors inevitably will lose their money. Decisions by lawmakers can strengthen or weaken various crucial parts of the innovation system that transforms an idea into a revenue-generating product. Thus, innovation journalists need to understand the markets and become well informed about the leading interests, their relations to each other, and their agendas.

This is the basis of the argument that innovation journalists may have a niche and an arena for a new professional community that could also become a new school of journalism. If the concept is viable, it might better equip reporters to contribute with dialogue and criticism that inspires companies and lawmakers towards sustainable growth, and scrutiny that inspires the public and investors to raise well founded questions.

The need for Journalism

Communities consist of people who interact around something they have in common. Community members share concepts that are parts of their identities, such as ideas, various aspects of professional and family life, forming a culture. A large part of the community awareness is the sharing of knowledge about the protagonists – who is who, who said what about who, and who is doing what.

In general, for most communities, if a statement is considered true or not by the community is not determined by how well the statement is formulated or by whom. It is determined by how often it is repeated and by how many sources. If many sources say the same thing often, people feel that is has to be true.

If a statement is well formulated, or stated by an authoritative protagonist, there can be an increased probability that it will be repeated by others, but usually active marketing is needed for a concept to reach "critical mass", when community members will start repeating the statement because everybody else is repeating it.

The free press is one of the dominant direct sources of shared knowledge in large populations today. While a teacher reaches hundreds of students, a journalist reaches at least thousands or millions of readers.

This is why it is essential for journalists to analyze trends and relations, call attention to key developments, kill rumours, blow the whistle on foul play, and call for the attention of the community. Furthermore, it is just as important that different journalists cover different viewpoints and perspectives, since objective truth does not exist and pluralism is the only constructive alternative. Most people repeat what is said in the news, so the journalists' level of understanding about the reported matters may often be of fundamental importance for the level of insight in the public debate and the quality of the shared common knowledge.

Most people don't have the time to carry out research on their own about issues they feel are of importance or are interested in. They rely largely on information produced by journalists. Four out of five Swedes read a daily newspaper about half an hour per day. The average Swede devotes six hours a day to media consumption including newspapers, magazines, books, radio, TV, CD, Video, and various Internet distributed content, according to statistics for 2002[v].

This makes the relation between the press and the community symbiotic, in some cases to the extent that they can define each other. The one helps the other to grow, and an open pluralistic community without its own press will be limited in important aspects that depend on the sharing of larger amounts of knowledge, such as economic growth.

Commercial publications depend on their ability to target communities successfully. Advertising brings in the money, and in order to sell ads, the publication needs to offer qualified target groups for sales ads or job ads. The price of ads depends on the access to readers and the amount and quality of the readers who are attractive for the purchasers of the ads.

Hypertext, multimedia and interactivity via the Internet offer publishers possibilities for deeper coverage than paper print can offer. Hopefully, once the Internet has become the leading distribution channel for news, we can expect intensified knowledge competition between news publishers. A high-quality news site with an archive and a search engine eventually becomes a great research library. Today it is more important than ever to invest in the development of journalistic education, research, and tools that promotes knowledge competition in the news media.

A few Swedish programs have already been created for promoting knowledge competition in the media, such as: basic training in Internet- and computer assisted research for professional journalists[vi], media training of academics heading collaboration projects between industry and academia[vii], and the development of advanced news delivery services and expert networks that enable reporters on deadline to find experts quickly[viii].

The need for Innovation Journalism

Innovation has a lot more to it than just technology. Innovation is also about management systems that drive growth. Peter Drucker[ix] defines innovation as "the act that endows resources with a new capacity to create wealth". For example, in the beginning of the nineteenth century there were dozens of harvesting machines on the market, but the farmers could not afford them. Eventually one of the many harvesting-machine inventors, Cyrus McCormick, invented instalment buying. This enabled the farmer to pay for a harvesting machine with future earnings rather than out of past savings. Suddenly the farmer could buy farm equipment.

Innovation is also about government management. An innovation-driven economy requires a country to improve its ability to compete and create a high living standard. According to Michael Porter[x], economic development is a process of successive upgrading, in which a nation's business environment evolves to support and encourage increasingly sophisticated and productive ways of competing for companies based there. A nation may upgrade itself from a factor-driven economy to an investment-driven economy to an innovation-driven economy, which is the most advanced stage of economical development. The national business environment in the highest stage is characterized by strengths in all relevant areas with a large degree of interaction in clusters. More sophisticated company strategies require, among other things, a highly skilled workforce, improved infrastructure, and more advanced research institutions. Sophisticated company strategies also require increased access to better information for the company's decision-making process.

Innovation is not only about technology and management; it is also about social systems and cultural trends. In 2002, in the sobering wake of the financial bubble often referred to as "the new economy", IBM chairman Lou Gerstner[xi] wrote that "technology itself isn't some force of nature that we simply direct or use. It, too, is the product of human intentionality and choice. So yes, we apply technology to solve customer problems. And we also apply marketplace knowledge to help shape our research agenda – whether it's the direction of the economy, or growth opportunities, or emerging forms of governance and education, or demographic and social trends, or discoveries in other fields such as life sciences."

Accordingly, in advanced economies today, public policy, higher education, R&D, manufacturing, marketing, and financing are being intertwined in innovation systems[xii]. The innovation-aimed interaction across industries, universities, and the rest of society is increasing together with the economic impact of innovation and the growing consciousness of innovation systems. "Innovation" is an increasingly popular expression, used in power contexts such as "innovation policy", which generates 33900 Google-hits. "Innovation policy" is thus gaining significance in comparison to the more established term "technology policy", which generates 399000 Google-hits[xiii]. This indicates a potential to establish innovation journalism.

Shared knowledge and increased interaction between different professions, such as engineers, business executives, academics, and politicians, strengthens the innovation systems. Many technologies are important for innovation, such as information technology, electronics, life sciences, materials, transportation technologies, etc. This is why we sense a growing need of news publications that can be shared across innovation systems, publications that offer a combination of what is offered today separately by trade journals, technology journals, business magazines, and newspapers. The potential readers are innovation professionals of various sorts, such as researchers, engineers, business executives, lawyers, public administrators, and politicians.

During the bubble of the new economy, those who handled the money took the technology for granted, and those who handled the technology took the money for granted. Everybody wanted to believe that the other actors were doing their jobs based on educated decisions and sound judgement. If the IT innovation systems had been investigated by qualified innovation journalists with a systematic approach, the chances would have been greater that the players within the new economy would have been forced to moderate their behaviour, perhaps decreasing the serious consequences.

Another significant factor that inflated the bubble of the new economy was the generally overenthusiastic estimation of the time needed for innovations to be accepted. Everett Rogers'[xiv] classical book "The Diffusion of Innovations" demonstrates that many innovations, even when they have obvious advantages, require lengthy periods - often many years - from when they become available to when they become widely adopted. Again, we would have gained from innovation journalism that could have blown the whistle. Ironically, skilled and consistent innovation journalism coverage might also have increased the speed of diffusion of innovations.

Innovators are exposed to social dangers because they create changes. The problem is not new, and Machiavelli[xv] explains it well: "the innovator has for enemies all those who have done well under the old conditions, and lukewarm defenders in those who may do well under the new. This coolness arises partly from fear of the opponents, who have the laws on their side, and partly from the disbelief of men, who do not readily believe in new things until they have had a long experience of them. Thus it happens that whenever those who are hostile have the opportunity to attack they do it like partisans, whilst the others defend lukewarmly". Innovation journalism can increase the chances for innovators to survive and succeed.

For innovation journalism to succeed, it is not enough that society needs it . Neither is it enough that many individual readers need it. Revenue streams are needed for innovation journalism to succeed. There has to be purchasers of ads, and innovation journalism offers a potential ad market, for example high quality job ads, ads for patenting services, ads for R&D tools, and services or ads for market research tools and services.

According to Porter, innovation is the top level of national competitiveness, so it seems reasonable that players on the innovation arena should be exposed to sufficient competition that can drive them to buy ads. It also seems reasonable that they should have sufficient financial resources to buy ads. Thus, one of the most important tasks ahead is to identify customers that are willing buy ads, and find the consumer statistics that will help the advertising departments to make a strong argument to sell ads.

As crass as it may sound to a journalist with a solid integrity, these are, nevertheless, the fundamental market mechanisms and rules by which their publications live and exist. However, it is not the journalists' task to make these considerations. Naturally, this task is to be independently performed by the commercial executives of the news operations as the reporting beat unfolds and develops.

The Internet increases the possibilities for niche marketing that ought to be particularly interesting for innovators. Consequently, there are enormously interesting possibilities for new services and drastically lowered marginal costs for production and distribution of news. However, the Internet also offers great challenges.

Testing if Innovation Journalism can become a discipline - a new reporting beat.

Sweden is the most knowledge-based economy in the world today, according to OECD[xvi]. Sweden spends 6.1% of its GDP on R&D, software, and higher education. Sweden is the cradle of well-known innovative technology companies such as Ericsson, ABB, Volvo, Saab, AstraZeneca, Pharmacia, and many others. It is also the cradle of other innovative companies such as IKEA and H&M. Through VINNOVA, the concept of innovation systems has penetrated society. The potential market for innovation journalism should be promising. There are some well-known trade magazines performing coverage that can be referred to as innovation journalism already. These publications are, consequently, positive to the concept. Furthermore, there are other business magazines and daily newspapers that have shown interest in the concept, thus willing to look closer into it.

But even though conditions look promising for the development of innovation journalism, initial government support is needed to get the development started. The Swedish media market has small resources for R&D and competence development, and very few philanthropic funding sources to confer with. The market is characterized by small margins, strong competition, and a high sensitivity to economic fluctuations. Media therefore focuses its coverage on the immediate future with small possibilities to invest in pre-competitive research.

It is important to stress that Innovation journalism, if it should be developed at all, must be developed by journalists. Journalists who may be called "innovation journalists" exist already today, but the question is if and when they will actually choose to call themselves "innovation journalists" instead of "business journalists", "technology journalists" or something else. And – if they do – who will dedicate himself to develop a professional community with his peers? What the nation can do is to give interested journalists and editors a chance to develop themselves together, maintaining respect for journalistic integrity at all times.

This is the challenge of the program "Innovation Journalism" [xvii]. The program aims at developing Innovation Journalism by sponsoring the personal competence development of some selected innovation journalists, encouraging team-building within the selected group, and supporting networking activities around the selected innovation journalists and the group.

VINNOVA is funding a fellowship of at least five innovation journalists. Two of the most important criteria for qualification is that applying selected reporters and editors self-identify as potential "innovation journalists" with full support from their highest newsroom executives. The selected fellows will be given an opportunity to work for at least four months at leading news, business, or technology publications outside Sweden[xviii]. The fellows will be given the opportunity to develop their skills within the beat of innovation journalism, and to extend their professional networks. The fellows will be encouraged to interact with each other during this time, perhaps forming the core of an innovation journalism community. A number of activities will be arranged to support the fellows, such as discussion groups, and visiting delegations with interested decision makers that may contribute in the creation of support networks around such a core community. This program model has been used earlier with great success.[xix]

The duration of the whole program is eighteen months. The program and is divided into three phases:

1. Preparations and selection processes (8 months)
2. International exchange for build-up of fellows' knowledge, skills, and international networks (4–6 months)
3. Workshops, possible pilot tests, and national networking (4 months).

One particularly important cornerstone of the program is journalistic integrity. The contacts between the selected fellows and the hosts are to be handled directly by each fellow. The program management will not determine how fellows and hosts pair up with each other. Neither will program management under any circumstance influence the choices of topics or the articles that are written by the fellows. A VINNOVA program committee, chaired by a formed editor-in-chief of a major Swedish newspaper, who will also be advising the program director throughout the process, performs the selection of fellows.

Development Communication, has been alternatively defined as a type of marketing and public opinion research that is used specifically to develop effective communication or as the use of communication to promote social development. Defined as the former, it often includes computerized linguistics analysis of verbatim responses to qualitative survey interviews and may, at times also involved consumer psychological "right brain" (emotional) research techniques. Defined at the latter, it refers to the practice of systematically applying the processes, strategies, and principles of communication to bring about positive social change. As most providers of *"communication development" research use proprietary approaches that cannot be elaborated upon without revealing proprietary trade secrets, the remainder of this article describes the latter definition.*[1] The practice of development communication can be traced back to efforts undertaken in various parts of the world during the 1940s, but the widespread application of the concept came about because of the problems that arose in the aftermath of World War II. The rise of the communication sciences in the 1950s saw a recognition of the field as an academic discipline, with Daniel Lerner, Wilbur Schramm, and Everett Rogers being the earliest influential advocates. The term "Development Communication" was first coined in 1972 by Nora C. Quebral, who defines the field as *"the art and science of human communication linked to a society's planned transformation from a state of poverty to one of dynamic socio-economic growth that makes for greater equity and the larger unfolding of individual potential."*[2]

The theory and practice of development communication continues to evolve today, with different approaches and perspectives unique to the varied development contexts the field has grown in.[3]

Development communication is characterized by conceptual flexibility and diversity of communication techniques used to address the problem. Some approaches in the "tool kit" of the field include: information dissemination and education, behavior change, social marketing, social mobilization, media advocacy, communication for social change, and participatory development communication.

History

The theories and practices of development communication sprang from the many challenges and opportunities that faced development oriented institutions in the last century. And since these institutions existed in different contexts, different schools of development communication have arisen in different places over time.[3]

Manyozo (2006) suggests that the history field can be broken down into those of six different schools of development communication, with the Bretton Woods school being the dominant paradigm in international literature, and the other schools being the Latin American, Indian, Los Baños, African,and the participatory development communication schools.

The growing interest for these kind of applications is also reflected in the work of the World Bank, which is very active in promoting this field through its Development Communication division and recently (June 2008) published the Development Communication Sourcebook, a resource addressing the history, concepts and practical applications in this discipline.

The Bretton Woods School

The "Bretton Woods school of development communication" is a term that has been applied to the development communication approaches that arose with the economic strategies outlined in the Marshall Plan after World War two, and the establishment of the Bretton Woods system and of the World Bank and the International Monetary Fund in 1944. The descriptive term is not widely used in the field, but has been used to differentiate between different "schools" or approaches to development which have historically evolved, sometimes independently, at later points in history and in other parts of the world.[4] Leading theorists under this school included Daniel Lerner, Wilbur Schramm, and Everett Rogers. Due to his pioneering influence in the field, Rogers has often been termed the "father of development communication."

Originally, the paradigm involved production and planting of development in indigenous and uncivilized societies. This western approach to development communication was criticized early on, especially by Latin American researchers such as Luis Ramiro Beltan and Alfonso Gumucio Dagon, because it tended to locate the problem in the underdeveloped nation rather than its unequal relations with powerful economies. There was also an assumption that Western models of industrial capitalism are appropriate for all parts of the world. Many projects for development communication failed to address the real underlying problems in poor countries such as lack of access to land, agricultural credits and fair market prices for products.

Failure of many development projects in the 1960s led to it reconceptualizing its top-down methods (Manyozo, 2006). The school has reviewed its approaches over the years and has been the most dynamic in testing and adopting new approaches and methodologies.[3]

The world bank currently defines development communication as the *"integration of strategic communication in development projects"* based on a clear understanding of **indigenous realities**.[4]

Institutions associated with the Bretton Woods school include:

- The United Nations Educational, Scientific and Cultural Organization (UNESCO),
- Food and Agriculture Organization of the United Nations (FAO),
- The Rockefeller Foundation,
- The Department for International Development of the United Kingdom, and
- The Ford Foundation.

Latin America

The Latin American School of Development traces its history back further than the Bretton Woods school, emerging in the 1940s with the efforts of Colombia's *Radio Sutatenza* and Bolivia's *Radios Mineras.* These stations were the first to use participatory and educational rural radio approaches to empowering the marginalized. In effect, they have since served as the earliest models for participatory broadcasting efforts around the world.

In the 1960s Paolo Freire's theories of critical pedagogy and Miguel Sabido's enter-educate method became important elements of the Latin American development communication scene.[5][6]

Other theorists who have influenced this school include Juan Diaz Bordenave, Luis Ramiro Beltran, and Alfonso Gumucio Dagron. (Mayonzo 2006, Mayonzo, 2005)

In the 1990s, technological advances facilitated social change and development - new media outlets began to emerge, cable TV signal coverage spread over more regions, and as the presence of communication firms grew so did an echoed global trend from major corporations.[7]

India

The history of organized development communication in India can be traced to rural radio broadcasts in the 1940s. As is logical, the broadcasts used indigenous languages such as Hindi, Marathi, Gujarati and Kannada.

Independent India's earliest organized experiments in development communication started with Community Development projects initiated by the union government in the 1950s. The government, guided by socialistic ideals of its constitution and the first generation of politicians, started massive developmental programmers throughout the country. While field publicity was given due importance for person-to-person communication - also because the level of literacy was very low in rural areas - radio played an equally important role in reaching messages to the masses. Universities and other educational institutions - especially the agricultural universities, through their extension networks - and international organizations under the UN umbrella carried the dev-comm experiments further.

Development communication in India, a country of sub-continental proportions, acquires many connotations. On one end of the spectrum are the tools and techniques locally applied by charitable and not-for-profit organizations with very close inter-personal relations among the communicators and on the other end is the generic, far-off, one-way sort of communication emanating from the government.

The need for development communication continues since a large population, over 600 million, lives in rural areas and depends directly on agriculture. Poverty is reducing as percentage of population but still over 200 million are very poor as of 2009. They all, and the urban slum dwellers, need government support in different forms. Therefore, communication from the government remains highly relevant. In addition to the traditional ways, a new form of communication is being tried by the union government to support its developmental activities, though at a limited scale. Called Public Information Campaigns, public shows are organized in remote areas where information on social and developmental schemes is given, seminars and workshops are held, villagers and their children are engaged in competitions, messages are given through entertainment shows. In addition, government organizations and corporate involved in rural businesses display their wares and services in stalls lining the main exhibition area. This approach brings various implementing agencies and service / goods providers while the information providers encourage the visitors to make the best use of various schemes and services available. Some state [=provincial] governments have also adopted this model to take their development schemes to the masses.

Community radio is another new medium getting a foothold in rural India, though in patches. NGOs and educational institutions are given license to set up a local community radio station to broadcast information, advisories and messages on developmental aspects. Participation of local community is encouraged. As community radio provides a platform to villagers to broadcast local issues, it has the potential to elicit positive action from local politicians and civil servants.

Africa

The African school of development communication sprang from the continent's post-colonial and communist movements in the late 1960s and early 1970s. Development communication in Anglophone Africa saw the use of Radio and theatre for community education, adult literacy, health and agricultural education.

In 1994 the FAO project "Communication for Development in Southern Africa" was a pioneer in supporting and enhancing development projects and programs through the use of participatory communication approaches. The FAO project, placed under SADC, developed an innovative methodology known as PRCA - Participatory Rural Communication Appraisal, which combined participatory tools and techniques with a strong communication focus needed to design strategies enhancing projects' results and sustainability. FAO and SADC published a handbook on PRCA and this methodology is still widely used today in various projects around the world.

Meanwhile, radio was being developed as a means of promoting rural development in Francophone Africa, with sponsorship from the Bretton Woods school institutions. (Kamlongera, 1983, Mlama, 1971, Mayonzo 2006, Mayonzo, 2005)

University of the Philippines Los Baños

The systematic study and practice of Development Communication in the Philippines began in the 1970s with the pioneering work of Nora C. Quebral, who, in 1972 became the first to come up with the term "Development Communication." In at least some circles within the field, it is Quarrel who is recognized as the "Mother" of Development Communication.

Quebral's work with the University of the Philippines Los Baños' Office of Extension and Publications evolved into today's College of Development Communication, which in 1971 became the first to offer degree programs at the Doctorate, Masteral's and Undergraduate degree levels.

Aspects of development communication which the CDC has extensively explored include Development Broadcasting and Telecommunications, Development Journalism, Educational Communication, Science Communication, Strategic Communication, and Health Communication.

Cybernetics approach

Another area of exploration for the CDC at UPLB is the aspect of development communication relating to the information sciences, the decision sciences, and the field of knowledge management. In 1993, as part of the then Institute of Development Communication's Faculty papers series, Alexander Flor wrote a paper on environmental communication that, among other things, proposed a definition of Development Communication expanded from the perspective of cybernetics and general systems theory[8]:

If information counters entropy and societal breakdown is a type of entropy, then there must be a specific type of information that counters societal entropy. The exchange of such information - be it at the individual, group, or societal level - is called development communication.

The Participatory Development Communication school

Focusing the involvement of the community in development efforts, and greatly influenced by Freirean critical pedagogy and by the Los Baños school, the evolution of the Participatory Development Communication School involved collaboration between First World and Third World development communication organisations. (Mayonzo 2006; Mayonzo, 2005; Besette, 2004).

Examples

One of the first examples of development communication was Farm Radio Forums in Canada. From 1941 to 1965 farmers met in groups each week to listen to special radio programs. There were also printed materials and prepared questions to encourage group discussion. At first this was a response to the Great Depression and the need for increased food production in World War II. But the Forums also dealt with social and economic issues. This model of adult education or distance education was later adopted in India and Ghana.

Similar to Canada's Farm Radio Forums, Radyo DZLB, the community broadcasting station of UPLB College of Development

Communication is a forerunner of school-on-air (SOA) concept that provides non-formal education for rural folk. DZLB has hosed SOAs on nutrition, pest management and cooperatives.[9] Currently, DZLB continues to air educational programming for farmers and cooperatives.

Instructional television was used in El Salvador during the 1970s to improve primary education. One of the problems was a lack of trained teachers. Teaching materials were also improved to make them more relevant. More children attended school and graduation rates increased. In this sense the project was a success. However, there were few jobs available in El Salvador for better-educated young people.

In the 1970s in Korea the Planned Parenthood Federation had succeed in lowering birth rates and improving life in villages such as Oryu Li. It mainly used interpersonal communication in women's clubs. The success in Oryu Li was not found in all villages. It had the advantage of several factors including a remarkable local woman leader and visits from the provincial governor.

A project of social marketing in Bolivia in the 1980s tried to get women in the Cochabamba Valley to use soybean recipes in their cooking. This was an attempt to deal with chronic malnurishment among children. The project used cooking demonstrations, posters and broadcasts on local commercial radio stations. Some people did try soybeans but the outcome of the project is unclear.

In 1999 the U.S. Government and D.C. Comics planned to distribute 600,000 comic books to children affected by the Kosovo War. The comic books are in Albanian and feature Superman and Wonder Woman. The aim is to teach children what to do when they find an unexploded land mine left over from Kosovo's civil war. The comic books instruct children not to touch the anti-personnel mines and not to move, but instead to call an adult for help. In spite of the 1997 Ottawa Treaty which attempts to ban land mines they continue to kill or injure 20,000 civilians each year around the world.

Since 2002, Journalists for Human Rights, a Canadian based NGO, has operated long term projects in Ghana, Sierra Leone, Liberia, and the DR Congo. jhr works directly with journalists, providing monthly workshops, student sessions, on the job training, and additional programs on a country by country basis.

Pressing issues within the field

Sustainable Development

Sustainable development is the process of maximizing the use of available resources in order to ensure the long-term wellbeing of present and future beneficiaries. Sustainable Development is a continuous progress which aims for and maintains a constructive state of living in society as preserved by social institutions and systems. However, sustainable development entails 1. Economic sustainability. 2. Social sustainability. 3. Cultural sustainability.

Community/People Participation

Community/People Participation is a voluntary involvement of an informed and motivated community while being equipped with proper knowledge and training in which they are equally gratified. It is the active involvement of members of a particular social unit in all aspects of developmental procedures (planning, decision-making, evaluating, monitoring, etc...). An equipped and facilitated environment is arranged to initiate the involvement of different willing social units, as to enhance the process of development and communication amongst its members. The participating people can inject or infuse to another people of the society largely, of the ideas and perception of the developmental process or project ongoing round about them. Community participation propels the objective of development communication. A variety of methodologies are used to advance people's participation, such as peer education, community mapping, participatory rural appraisal, among others. The sole heart of community participation is the people.

Complications in practice

Like any intervention covering the society, and more because human communication itself is complicated, development communication can become complicated. The complications increase when we deal with diverse societies over a large area, such as India; when we try to change behavior of the recipients and in the process bring in many types of media / persuasive skills; and when the ground realities do not allow results to reach the target audience, i.e. there is disconnect between the ground realities and messages. Cultural factors including local rituals and mores, nuances of language, gender perceptions, affect the reception of messages and their impact. The success of dev-comm. will, therefore, depend on the credibility of the messenger, the simplicity and directness of the message, and its location-specificity. This aspect assumes more significance when we communicate in complex societies.

Other related definitions

Development Communication is recognizing the power of communication as a catalyst for social development. It is also the utilization of existent communication tools and applicable theories for result-driven strategies for the advancement of society.[

Development Communication can also be defined as purposive communication intended for a specific target audience that allows for the translation of information into action resulting in a higher quality of life.

It is greatly linked with the concepts of Sustainable Development (which can be defined as the improvement of a community using information and technology and the community's ability to maintain the created ideal state without compromising its environment and resources). It also relies greatly on Community and People Participation, which is the voluntary involvement of a group of people in a development activity with full knowledge of its purpose that will allow them to grow individually and as a community.

Development communication is the process of eliciting positive change (social, political, economic, moral, environmental, etc.) through an effective exchange of pertinent information in order to induce people to action.[If the present understanding of communication and development are integrated, the horizon of a practitioner's understanding of development communication will widen. It will not be limited by historical definitions but include the following elements and more such aspects. Development "communication" thus would include: information dissemination on developmental schemes/ projects, communication for eliciting positive change, interactivity, feedback on developmental issues, feedback/reverse communication for eliciting change. On *development side,* sustainability issues need to be given proper importance vis-a-vis economic development.

Development Support Communication:- The term can be described as development planning and implementation in which adequate action is taken of human behavioral factors in the design of development project and their objectivities. It addresses development planning and the plan of operation for implementation. Development support communication is urgently suggested by UNESCO, UNDP and communication scholars and practitioners worldwide. It (DSC) stands for linking all agencies involved in the planned development works such as political executives, political planners, development administrators, subject specialists, field workers, opinion leaders, media representatives, researchers and the beneficiaries who continue the final delivery points and the consumers of the information. The route of communication envisaged are not only vertical as flowing from upper level to bottom or bottom level to upwards but also horizontal between the institutions and personnel connected with the process of development.

Objectivity

The fundamental objective of DSC,is to communicate the latest skills, knowledge and innovation to the agriculturists so that by adopting them the agriculturists may increase their output manifold. In this connection three vital groups are identified which are as follows:

(1) Innovation or Knowledge generation. (2) The political leaders or government of the state. (3) Users of the knowledge or agriculturists. A very close interaction is necessary among the three groups as mentioned above, to achieve the success of development support communication (Phazcom 26.02.09).

International Communication - the intellectual field that deals with issues of mass communication at a global level - is sometimes also called development communication. This field includes the history of the telegraph, submarine communication cables, shortwave or international broadcasting, satellite television, and global flows of mass media.

❑❑❑

Chapter 6

The Concepts of ICTs in Modern Journalism

The Concept of ICTs in Modern Journalism

The role of the journalist as a communicator

Basic concepts of communication -

Encoder— —Medium— —Decoder

The Concept of ICTs

ICTs are basically information handling tools - a varied set of goods, application and services that are used to produce, store, process, distribute and exchange information.

They include *radio, scanners, photocopiers, cell phones, digital cameras, recorders (video, audio etc.), television and telephone, computers and the internet as well as satellite and wireless technology.*With appropriate content and applications the ICT tools combine to form a "network world". The network size may vary depending on the number and type of tools involved. ICT was coined to reflect the seamless convergence of digital processing and telecommunications which include hardware, processes and systems that are used for storing, managing, communicating and sharing information. This could be seen as an integrated technology.

ICTs therefore represent the different technologies involved.

The significant role of the Journalist

- Uses ICTs to foster the processing, and dissemination of information and knowledge by separating content from its physical location
- Uses information to break boundaries-geographical, cultural thus making it possible for the resultant interconnected digital world to herald the fluid and seamless flow of information in McLuhan's coinage of the global village.

Benefits of the use of ICTs in Modern Journalism

- Discuss the challenges that have confronted the journalist in the use of obsolete tools or old technologies.
- Discuss how the journalist stands to gain in the use of ICTs today
- Benefits of the use of ICTs in Modern Journalism

ICTs offer the developing world the opportunity to leapfrog several stages of development

- Timeliness (real-time) of communication.
- The global village concept becomes real.
- Confidence and reliability of communicated material.

Specific advantages

- Access to information for private and professional decision making. Prices of Market/Farm goods.
- Reduction of space and distance as a factor for social and economic participation.
- Access to opportunities- ICTs empower SMEs, local and community groups , women, marginalised and disenfranchised groups.
- Greater ability to learn

- Greater environmental awareness - crop failures, preparation of emergency reliefs etc
- More awareness factors
- Greater ability to influence and participate in decision-making.
- Speeding up transactions
- Improved trade, commerce enabled
- ICTs have been used to help fishermen, farmers for schools of fish.
- Used to pick satellite imagery of land-cover
- ICTs and the Journalists role in Human development
- People do not need the tools of the modern Journalism to survive.
- People want efficient ways of sharing information about livelihoods and employment.
- The desired information is best appreciated when modern ICTs are used. *ie when issues are current and there are opportunities to get involved.*
- There is an interplay of people-centeredness and ICTs in the journalist's work.

Group Work

- Be in a group of four.
- Discuss three core functions in your work as a Journalist/media professional. How can you make use of use ICTs in carrying out each of those functions?
- List ICTs necessary for carrying out your work. Which of them are not easily accessible to you at your place of work?
- How do people benefit from having access to ICTs? State the benefits you are likely to derive from easy access to ICTs and possible recommendations (only two) you can give to your employer.
- for your

Role of ICT

Farming Information and awareness package through print, mass and now emails is possible. The role of impeditive forming, economic survey and evaluation of farming and women's integration need attention. Women, the half of agricultural work overcome by educating them through TV, cassettes, e-mails or other modern earn indication

Appliances

Extension provides refined tech. Participatory research and educational modules to bridge the productivity gaps.

Future growth in agriculture would have necessarily come from increased productivity from a shrinking natural resources base through efficient management.

ICT: - Barriers of ICT Concepts

Applications of ICT for argil/Development

1. GIS
2. Remote Sensing
3. E governance
4. Cyber Extension
   - Participation
   - Problem Solving
   - TOT
   - Innovation
   - Group formation

Role of ICT in agriculture development

1. *Economic development*: Application in planning and market information for agriculture procedures.
2. Community Development Applications.
3. Research/Education Applications
4. Small and medium enterprise development
5. New media networks
6. Providing decision support to public administrates for improving planning and monitoring of development programmers.

7. Improve services to citizen, and bring in transparence.
8. Empower citizens through access to information and knowledge
9. Help in training to improve the functioning of developmental organization and expand employment opportunities.
10. Market
11. Linkage
12. Services
13. Small and median former

Digital divides: Distance and difference caused among because of the haves and have not's of the ICT.

Impact: Hurdle in conformity in gains of the ICT

Two biggest hurdles

1. Access and content
2. Linguistic and technological divide

Each medium of communication is dominated by specific linguistic traditions. This limits the access to the medium as terms of understanding and utilizing the medium.

In case of ICT, two kinds of divides are prominent

(i) A divide between English and non English speaking communities where comprehensible content is available only to a selected few who have knowledge of English.

North south divide : Between developed and developing countries.

National level initiative in ICT for Agricultural Development

1. Gyandoot project
2. Warna wired village project
3. Information village project of MSSRF
4. I kisan project of Nagarjana group of companies (Andhra Pradesh)
5. Milk collection centers of dairy cooperative of (AMUL) Gujarat.

6. Tarahoat.com by development alternatives (Uttar Pradesh and Punjab)
7. Land Record computerization (Bhoomi)
8. Knowledge network for grass root innovations - SRISTI (Gujarat)
9. Mahiti Samutra (Karnataka)
10. Centre for alternative agriculture media (CAAM)

Extension Teaching Methods

Define: Devices use to create situations in which communication can take place between the instructor and the learner.

Functions

1. To provide communication so that the learner may see, bear and do the things to be learnt.
2. To provide stimulation that causes the desired mental and/or physical action on the part of the learner.
3. In brief, to take the learner through one or more steps of the teaching - learning process, viz attention, interest, desire, conviction, action and satisfaction.

Classification

A. According to use

i. Individual contact
ii. Farm and home visit
iii. Office calls
iv. Personal letters
v. Result demonstrations

B. Group contacts

i. Method demonstration meetings
ii. Leader training meetings
iii. Lecture meetings
iv. Conference and discussion meetings
v. Meetings at result demonstrations
vi. Tours
vii. Schools
viii. Miscellaneous meetings

C. Mass contact

i. Bulletins
ii. Leaflets
iii. Necos stories
iv. Circular letters
v. Radio
vi. Television
vii. Exhibits
viii. Posters

2. According to form

a. Written

i. Bulletins
ii. Leaflets
iii. News articles
iv. Personal letters
v. Circular letters

b. Spoken

i. General and special meetings of all kinds.
ii. Farm and home visits
iii. Office calls
iv. Telephone calls
v. Radio

c. Visual or objective

i. Result demonstrations
ii. Exhibits
iii. Posters
iv. Motion pictures, charts, slides, and other visual aids

d. Spoken and visual

i. Method demonstration meetings
ii. Meetings at result demonstration
iii. Meetings involving motion pictures charts and others visual aids.
iv. Television

Selection of extension teaching methods

Factors to be considered in selection and combination of extension methods:

1. The audience
   a. Individual and collective differences
   b. Size of audience
2. The teaching objectives - Thinking attitude/feeling action/ skill
3. The subject matter
4. The stage of development of extension organization
5. Size of extension staff
6. The availability of certain communication media.
7. The relative cost of the method.
8. An extension worker's familiarity.
9. The nature and complexity of the subject matter.
10. The extent to which people already know and are practicing the recommended practice.
11. The relative economic and social importance of the behavioral changes desired.
12. The amount of extension worked efforts likely to be devoted to a particular phase of a program me
13. The kind and extent of assistance that may be provided by local leaders and others.
14. The member in which the methods selected complement each other in attaining the behavioral changed.

Audio - Visual aids

According to **Haas and Packer (1964)** "an audio visual aid is an instructional device that can be heard and seen"

According to date

All materials used in the classroom or other teaching situations to facilitate the understanding of written or spoken words.

Classification

Evaluation

Senses involved

According to evolution

1. First generation media: Handmade charts, graphs, exhibits.
2. Models, hand written material etc
3. Second - Generation media: Printed/Illustrated tects, printed graphics, work books
4. Third generation: Photographs, slides film strips, films recordings, radio tele communication.
5. Fourth generation: Television, programmed instruction, language laboratories, electronic digital computers.

According to senses involved

1. Audio Aids

1. Radio talk
2. Tape recording
3. Disc recording

2. Audio - Visual Aids

1. Motion pictures
2. Telecasts
3. Video recording
4. Sound synchronized slides
5. Radio vision

3. Visual

a. Projected

Films, slide, OHP, Model, specimen puppet, real objects, 3d films

b. Non - Projected aids

i. Presentation

1. Flash cards
2. Flannel graphs
3. Charts
4. Flip books
5. Maps

ii. Display

1. Bulletin boards
2. Photographs
3. Posters
4. Wall charts
5. Exhibits

iii. Literature

1. Illustrated
2. Leaflets
3. Comics

According to contribution to learning

1. According to edger date

- Selection of audio - visual aids:-
- Factors influencing selection of audio - visual aids:-
- Audio - visual aids are used singly or in combination, taking the following factors into consideration:
    - *The teaching objective* i.e. the type of behavior change you want to bring about gaining information, or changing attitudes, or learning some skill.
    - The nature of subject matter being taught.
    - The nature of audience age level education level, interest experience, knowledge of subject
    - *The size of audience*: Flash cards can be used for a small audience only, motion picture, for a large audience.
    - *Relative cost of the various aids* - effective aids needs not necessarily be expensive
- The teacher
    - The extension worker's familiarity with and skill in using the several aids.
    - This originality and skill in selection, preparation and use of aids.

- *The availability* : An effective extension worker makes use of indigenious materials, when the teaching aid he would like to use.

Evaluation of audio - visual materials

1. Do the materials give a true picture of the idea they presents.
2. Do they contribute meaningful content to the topic under study?
3. Is the material appropriate for the age intelligence and experience of the learners?
4. Is the physical condition of the materials satisfactory?
5. Do they make learner better thinkers, critical minded?
6. Do they tend to improve human relations?
7. Is the material worth the time and effort involved?

The effective use of A.V. aids

1. Planning
2. Preparation
3. Presentation
4. Evaluation

Planning

- Know clearly the objective of presentation
- Plan well in advance
- Anticipate size of audience so closely as possible
- Plan for the use of a variety of colorful visual aids.
- Determine the appropriate timing for the presentation

Preparation

1. Prepare by refreshing or previewing in order to make a smooth presentation.
2. Select as convenient and as comfortable a meeting place as possible.
3. Anticipate need for special lighting or for a lately darkness.
4. Make sure that all equipment is in good working order.

5. Arrange the audio - visual aids in sequence.
6. Keep aids out of sight until actually required for use.

Presentation

1. Motivate the audience and stress the key points they should observe during presentation.
2. Present aids at the right moment and in proper sequence.
3. Display only one aid at a time.
4. Remove all unrelated material.
5. Stand beside the aid, not in front of it.
6. Speak facing the audience and not the aids.

Evaluation

1. In the end evaluate by providing for discussion and application to dispel and discover misunderstanding if any.
2. Undertake follow-up studies and observe results.

Research finding about the uses or merits of A.V. aids

Research has shown that audio - visual aids help

1. The learner to learn faster, more thoroughly remember longer
2. The teacher to organize his teaching material in a systematic order.
3. Clarify ideas being presented
4. Impress ideas more indelibly on the mind vitalize and make teaching more real.
5. Picture experiences outside one's own experience.
6. Combat verbalism
7. Overcame the language barrier
8. Attract and and attention
9. Arouse and sustain interest
10. Stimulate thinking and vitiate action.
11. Change attitude or paint of view.
12. Save time because they intake learning easier and faster.

Disadvantages

1. Learner may sometimes form mistakes and distracted impression.
2. Temptation for the teacher to narrow down his teaching to only a few big ideas.
3. Spectatorism risk.

Theories of communication distortion

1. Theory of communication distortion. States that larger the volumes of communication, larger would be the communication distortion.
2. Theory of communication distortion in transit. Emphasizes that message pass through many layers in an organization therefore repetition and reproduction of the same message is multiplied into many reduces the efficiency of communication.

Theories of communication

1. Linguistic theory: Deals with the description of a language system.
2. Psycholinguistic theory: Focuses attention the behavior of an individual in the use of language.
3. Socio-linguistic theory: Explains patterns of variation in language across different people, stimuli relation shop, context and competence.

Theories of interpersonal communications

1. Kelnan paradigm
2. Bauers scheme
3. Festinger scheme
4. Attribution theory
5. Game theory
6. Social comparison theory
7. New comb's theory
8. Stimulus response theory
9. Linguistic theory
10. Psycho linguistic theory
11. Socio linguistic theory

In general the theory of interpersonal communication explains it as the face to face interaction between people who are consistently aware of each other. Here the mutual presence of the message generator and receiver is required. Each person assures the rules of sender and receiver of message who are independent and interchangeable. The behavior of participants in interpersonal communication are so intertwined in the process that it is difficult to separate sender from receiver.

Attribute theory: Gives stress to the psychological state, motives and intentions of the individuals involved in communication.

Game theory: View interpersonal communication as the process of transaction between the people to balance possible gains and losses.

Social Comparison theory: Persons communications in order to verify their perceptions

Of the world assumption about themselves.

New comp's theory: Explains the thinking process of the who are reaching to one another simultaneously.

Networking system of information

A networking is the means by which computers store and exchange information and resources access either short distance (LAN) or globally (wide area network) making best use of networking technology is about optimum use of business resources, enhancing productivity and efficiency, reducing coats and gaining competitive advantages.

Teleconferencing

The new integrated digital communication system is being used. This technique will add new dimensions of control timing and flexibility to communication. Instead of traveling miles for a meeting conference rooms will be wired and can eras will be used to carry visuals.

Advantages

1. Time and cost for traditional day to day communication will thus be reduced.
2. It way for international integration.
3. It supports to provide/exchange latest agricultural technologies available in different countries for the farming communities.

Two types

(i) Audio conferencing

(ii) Video conferencing

Problems of ICT

1. Infrastructure barrier
2. Illiteracy
3. Poverty
4. Policy hurdles
5. Low people participation
6. Lack of will to implementation
7. Heterogeneous adoption
8. Digital divide

Prospects of ICT

1. Marketing information
2. Remote sensing
3. DSS
4. Realizing the knowledge potentials
5. Revival of extension organizations
6. Convergence with agricultural extension
7. Meeting the challenges of globalization
8. Online services - information, education training, monitoring, evaluation.
9. E-commerce
10. R-E-F interaction
11. Question - answer service
12. E-governance
13. Up to date information service
14. Database creation
15. Predictive in for - diseases, pest, weather
16. Land record facilitation

Interpersonal communication methods : Basis and key for interaction with others.

Communication takes place in following ways

Criteria	Mass Media	Interpersonal
1. The senses they stimulate at a time	Limited to one or two	All senses
2. Opportunity for feed back	Minimum delayed	Maximum immediate
3. Control of the pace	Uncontrollable	Controllable
4. Message codes	Highly verbal verbal	High Non verbal and
5. Multiplicative power	High	Low
6. Direction of message flow	One way	Two way
7. Message accuracy	High	Low
8. Power to preserve message	High	Low
9. Ability to select receiver	Low	High
10. Ability to overcome selectivity and noise	Low	High
11. Ability to meet specialized need	Serves common needs	Serves specialized needs
12. Speed to a large audience	Fast	Slow
13. Possible effects	Knowledge gains change	Attitude and action
14. Neutrality to type of receiver	Not neutral need literate audience or homogenous group	Any type of audience or people

1. Intrapersonal - talking with self, as in thinking
2. Interpersonal - Face to Face communication between two or more people
3. Mass Communication - The use of print, radio, TV and other forms of media in addressing a public 'masses.

Interpersonal Communication : Refers to that situation where the source and receiver can see each other, talk back and forth, interrupt each other, and make responses which both can readily observe.

Non interpersonal : Refers to that particular situation where something (it may be space, time or physical barrier of some sort) is placed between two interacting individuals.

Difference between interpersonal and Mass Communication

Ingredients/elements of interpersonal communication

1. Effective listening
   - Wishful thinking
   - Attention
   - Semantic
   - Talk person to person
   - Attitude
   - Get the fact
   - Excessive talking
2. *Empathy*: Sensitivity
3. Self awareness
4. Memory - Short term long term
5. Handling people

Making an impression to others

- ❐ Posture
- ❐ Look, listen and learn
- ❐ Expression
- ❐ Appearance
- ❐ Speech
- ❐ Eagerness to help

Consequences of poor interpersonal communication skills

1. Stress
2. Look of communication
3. Irritation
4. Close mindedness
5. No team sprit
6. Lack of credibility

7. Poor self esteem
8. Suspicion
9. Loss of productivity
10. Isolation
11. Poor health

Methods of communication interpersonal

i. Official/formal

1. Radio
2. Doordarshan
3. Newsletter
4. Exhibition
5. Bulletins
6. Tour
7. Written material
8. Poster
9. Meetings
10. Agricultural Unit
11. Local authorities

ii. Non official/Informal

1. Common talk
2. Members of the family
3. Neighbors and friends
4. Shopkeepers and leaders

Importance of interpersonal communicator

1. Important in individual contact method.
2. Provide Homophiles stale of communication
3. Bridge the communication gap.

Communication methods: Oral, written, information technology based.

Empathy

Berlo 1960 "Empathy is the process through which we arrive at expectations, anticipations of the internal psychological states of man"

Theories of empathy:

i. Inference theory:- Each communicator develops a concept of self in a communication situation. He can observe his own physical behaviouir directly and corelate his behavior symbolically to his own internal psychological status, feeling, emotions, thoughts etc on the basis of his prior interpretations of self. He makes inferences about the internal states of others.

ii. Role - taking theory (Mead 1937):- This theory developed by Mead (1937) argues that the concept of self does not precede communication, rather it is developed through communication. In the case of a child, for example, the infant actually plays other's rates without interpretation. he imitates the behavior of others. Later, on individual plays other people's rate with understanding still later. No single theory is sufficient to explain the legality. Generally, human beings utilize both these approaches of empathy for the purpose of communicating with others.

Empathic accuracy

1. As group increases, empathic accuracy decreases
2. When prior communication is minimal empathic accuracy is less
3. When we are insensitive to the behavior of others empathic accuracy is less.
4. Empathic accuracy also decreases when we are not motivated in a communication encounter.

"So a knowledge of the composition and working of a social system is useful in making predictions about new members of that systems will behave in a given communication situation".

Interaction : Term interaction refers to mutual Berlo, "1960" and reciprocal influencing of each other's behavior. The term interaction names the process of reciprocal role taking, the mutual performance of empathic behaviors. If two individuals make inferences about their own rules and take the role of others same time, and if their communication behavior depends on reciprocal taking of roles then they are communicating by interacting with each other.

All communication is not interactional; however, to the extent we are in an interactional situation, our effectiveness, our ability to affect and to be affected by others, increases. Thus, 'we can define interaction as the ideal of communication, the goal of human communication' "Berlo, 1960".

Heterophony/Homaphily

According to Rogers and shoemaker (1971)

Heterophony is the degree to which pairs of individuals who interact differ in certain attributes, such as beliefs, values, education, social status and so forth.

Effective communication

Relationship in communication

1. Physical interdependence/feedback
2. Action-reaction interdependence
3. Empathy
4. Interaction, and
5. Heterophony etc

Social Networks in communication

Configuration of network has a direct effect on communication behavior.

Star

1. Four person network
2. Five person networks

Mass communication: Communication which involve large, heterogeneous and anonymous audience"

Nature of mass audience Directed towards relatively large, heterogeneous and anonymous audience rapid, public and transient heterogeneous audience

- Large
- Heterogeneous
- Anonymous

Nature of Mass Communication

1. It is directed towards a relatively large, heterogeneous and anonymous audience.
2. Messages are transmitted publically.
3. It is often timed to reach most audience members simultaneously.
4. It is transient in characters.
5. The communicator tends to operate within a complex organization
6. It may involve great expense.
7. It was a mechanism that is capable of reproducing a large no. of identical messages, almost simultaneously and is able to transmit within a relatively short period to an audience that may even be physically separated in terms of space.
8. These are new technically, based geographical areas which provides, bichologly and informech an the new innovation and training and material related to them.

Technology Parks/Rural Tech Parks

Also known as science parks agricultural technology parks

Main objective : To promote Agricultural food processing, Industries in clusters in areas where there is predominant production of processible argil/hurt products.

Specific objectives

Quick information dissemination for farmer's school children and general public A place to see and learn tatest recommendation of crop cultivation and technologies generates by both public and private institutions.

1. A place to see and study heritage agriculture system and future high tech agriculture. Education and training mechanism for village level extension workers as face to face training of widely dispersed large number is practically impossible task.
2. Establishment of live agriculture crop museum.
3. Information network among agriculture officers, government and agriculture institutes.

4. Aesthetically beautiful landscaping
5. A site for eco tourism/Agro tourism
6. Generation of funds within the system for its sustainability say sustainer
7. One stop shop - sale centre for plants, seeds and other products of government institution and the private sector entrepreneurs of instructional media material information centre etc.

Main elements of A Parks/T. Parks

1. Venue: - Should be at the site of majority of argil related institution
2. Main sections and presentations rice, section, vegetable garden, roots and laber garden, fruit garden, dry zone crops garden
3. Exhibition stalls or main crop categories
4. Nursery for demonstration and sales
5. Mushroom house
6. Food tech. demonstration
7. Floriculture demonstration and sales area for private sector
8. A site to demonstrate traditions/indigenous agriculture in India - again agriculture.
9. A site to demonstrate future high tech agriculture
10. Area of student for discovery learning
11. Meteorology
12. Animal production and heath
13. Export argil crops
14. Sales centre
15. Information centre
16. Food courts
17. Resting areas

Policing initiative: Developing of world-class infrastructure for agro food processing industries through establishment of agro food tech parks.

These are of the state technology furnished locations wait provide info, being, R & D and other services related to new Agriculture and

Horticulture forms which are located within the radius of 80.100 Km from the proposed food agro technology parks would be linked to these parks.

One of the major thrust given by of India for development of world infrastructure for Agro food industries is establishment of "Agro Food Technology Parks" for which financial assistance of Rs 4 Crore per park is available from GoI.

Common facilities and infrastruce: R & D, were housing, cold storage, quality assurance labs, common effluent treatment plants.

Advantages

- RPT - Proposal to establish
- RPP initiated in 1992
- Debated over foodies
- 1995 Min of RD accepted proposal
- 10 crore badge estimates area submitted in place.
- Collorborative offer betweenCAPART

Scope for TOT through Tech demo Key communication - opinion leader

"Key comm. or OL are persons in any social system who are sought out for information and advice on general or specific topics"

Rogers and Kincaid (1981) define opinion leadership as the degrees to which on individual is able to informally influence others knowledge attitude, or overt behavior in a desired way with relative frequency"

Concept was first developed by **Lazars field (1944)**

Identification methods

i. Sociometric method
ii. Self designation technique
iii. Information ratings

Characteristics

i. External communication
ii. Accessibility
iii. Social status
iv. Innovativeness
v. Relationship to the social systems

Role

1. Indirect communication of innovations in a social system
2. Legitimization of new ideas for diffusion into the system
3. Self adoption of a relatively greater number of innovations
4. Help in securing services and supplies to the members in a social system

Characteristics affecting their role

i. Social participation
ii. Education
iii. Value orientation
iv. Socio-economic status

ATIC: "Agriculture Technology Information Centre".

The ATIC is a "single window' support system linking the various units of a research institution with intermediary users and end users (farmers) in decision making and problem solving exercise".

The rationale for establishment of ATIC are:

1. To provide diagnostic service for soil and water testing, plant and livestock health
2. To supply research products such as seeds and other planting materials
3. Providing information through published literature and communication material
4. Providing an opportunity to the institutes, SAUS to generate some resources through the sale of their technologies

The important criteria for agriculture technology information are:

1. Availability of new technologies
2. Relevance of new technologies
3. Responsiveness of new technologies
4. Sustainability of such unit within the overall institutional system

Target/users of info

Beneficiaries of ATIC

1. Farmers
2. Farmers - entrepreneurs
3. Extension workers and development
4. NGO's and
5. Private sector organizations

Objectives

1. To provide a single window delivery for producers and services.
2. To facilitate/dissect the farmers access to the institutional resources.
3. To provide mechanism for feedback from the users to the institute.

The information services and supply include :

1. Soil and water sample testing facility
2. Plant clinic and diagnostic centre
3. Rhizobium culture
4. IPM - organic and biopesticide/NPY
5. Seed and planting material, small implements
6. Fertilizer quality testing
7. Tissue culture and plant materials
8. Farm literature - leaflets, pamphlets, journals
9. Audio, video cassette on crops
10. Video show
11. Processed products - cereals milk, meat job
12. Cafeteria
13. Technology park (display) exhibition
14. Veterinary animal clinical services
15. Poultry strains, livestock breeds, seed etc.

Management structures of ATIC

PMC

TDMC

PIU DDGAG Extn - TDU

Zonal coordinator

Director of Institute/Director of extension of SAU

ATIC Management committee

E governance: E governance is defined by M Bakes (Bakus, M. 2001), "Application of electronic means in

a. Interaction between government and citizen and between government and business as well as.
b. In internal government options to simplify and improve democratic, government and business aspects of governance."

"E-governance is about transforming the relationship between governments and their citizens through the use of information technologies".

Benefits - Direct

- Faster, easier more convenient service convenience, near to home, more services at one place.
- Faster processing shorter wait shorter queries at government offices.
- Less number of trips to government officer.
- Better interaction environment no harassment.
- Reduced transport cost.
- Avoids wage loss for daily wage earners.
- Better quality service.
- More accurate and legible documents, easy recovery from errors.
- Lesser corruption increased transparency.
- Improved access to offices.

Indirect/Derived

1. Improved complaint handling, complaint tracking.
2. Introduces transparency in data, decision/actions, rules procedures and performance of government agencies.
3. Automated processes to take away discretion.
4. Reduced effectiveness in decision making.
5. Entry point for simplification g rules and engineering process.
6. Makes decision traceable takes actions.
7. Builds accountability.
8. Documentation to citizens to follow-up.
9. Tracks procedural, human delay.
10. Easier to fix responsibility accountability.
11. Modularize tasks making outsourcing possible.
12. Introduces completion among delivery channels and departments.
13. Standardized documentation of common of objects leads to effective supervision through comparative indicators.
14. Centralizes data for better audit and analysis integration of data across application.

Critical success factors

1. Political will
2. Overall vision and national (e-strategy)
3. Full ownership and support of it
4. Effective and appropriate project
5. Management
6. Access
7. Education of clients
8. Participation of clients
9. Culture
10. Cooperation and collaboration with private sector

11. Technology
12. Human Resources
13. Legal framework
14. Finances

Components of E-governance

1. Political
2. Technological
3. Social
4. Cultural
5. Psychological
6. Service

Interaction among/or ecology of E-governance various components

Stages of E governance

Degree of progress made by government towards the ultimate e-governance goal.

a. Presence
b. Interaction
c. Transaction
d. Transformation

According to layne on lee (2001)

1. Cataloguing
2. Transaction
3. Vertical integration
4. Horizontal integration

Cyber extension : Initiatives in India and other developing countries cyber extension can be defined as the extension over cyber space'. As the word extension is subjected neutral, so is cyber extension. But in the applied context of agriculture, cyber extension means". Using the power of online computer networks with the help of communication channels to deliver the content in the form of text, graphic, audio and video either passively or interactively to facilitate dissemination of agricultural technology".

Cyber extension tools

1. E-mail
2. Interactive export systems on crop
3. pest and disease
4. Internet processing for extension information
5. Video conferencing
6. Call centers, SATCOM Networks
7. Discussion groups and news groups

The broad classification of ICT initiatives currently in place :

1. Central government initiatives to provide connectivity to extension system NATP-ITD component, ICAR-extent component under NATP, MIC initiative in Kolhapur, Sangli district of Maharashtra.
2. State initiative MH government to promote info kiosk, Kerala government Akshaya project to connect village panchayats.
   1. Centre
   2. State
   3. NGO
   4. Private
   5. Pure information
   6. Multiple rebel service
   7. Knowledge networks
   8. NGOs project: information village project in Pondicherry by MSSRF Chennai. Rural connectivity and allied services providers Drishtee, N Logue
   9. Private and cooperative sector transaction related initiatives such as there of ITC, Mahindra, EDD
   10. Pure information services providers such as Ikisan, community India, Agriwatch, India agro net etc.
   11. Multiple rural services which include agriculutral extension such as WARNA, Wyandot, Taranaat etc.

12. Multiple rural services which include agricultural extension
13. Knowledge networks and knowledge banks, Honey Bee Network, Harit Gyan, Indian society of AB professional detailing.

CE current benefits

- CE is now an established complementary and supplementary of mechanism of argil extension. All Indian State Agriculture university (IGNOU) included cyber extension in Agriculture extension as an integral part of their undergraduate and post graduate curriculum.
- The GoI indicate to promote the of cyber extension in the reform of extension system.
- The launch of KCC (Kisan Call) has added a new dimension in to the Indian extension system.
- It can be safely concluded that the Indian agriculture is on threshold of a new communication mechanism.
- Cyber extension will further improve the international competitiveness and penetration of Indian agricultural products, by providing timely information across to farming community of supply side and the prospective consumers on the demand side.

Cyber extension

The process

- Develop state-of-the-art ICT infrastructure of connect key stack holders.
- Creating ICT awareness in all development departments.
- Create information packaging mechanism at keg participating agencies.
- Network with e.g. initiatives of concerned state/district.
- Create a model cell in each state to monitor the progress of CE.
- Identify a nodal connectivity agency for "cyber extension" in India.

- Services provided/benefits of CE
  - Providing interaction among SEF and other rural people.
  - Providing upto date news and information service.
  - A question and answer service.
  - Creation and maintenance of statistical database.
  - Providing the details of poverty all aviation schemes on the internal.
  - Providing states of various govt. programmers' and details about their implementation mechanism.
  - Hosting website by major institutions participating in the agricultural extension.
  - Launching online rural development and extension journals.
  - Providing internet access at district and block level agricultural and rural development officers.
  - Opening of cyber cafes to enable educated rural people and extension workers.
  - Providing maps that display different features such as pop density crops planted.
  - Providing video clips to demonstrate complex procedures.
  - Providing mechanism of user/beneficiary feedback for the public sector scheme.

E-learning: Electronic learning is a type of technology supported education/learning (TSL) where the medium of instruction is through computer technology, particularly involving digital technologies. E-learning has been defined as pedagogy empowered by digital technology. In some instances no face-to-face interaction takes place.

E-learning services have evolved since computers were first used in education there is a trend to move toward included learning services, where computer based activities are integrated with practical or classroom situation.

Pedagogical approaches or perspectives

It is possible to use various pedagogical approaches for e-learning which includes instructional design. The traditional pedagogy of instruction which is curriculum focused and is developed by a centralized educating group or a single teacher.

1. Social Constructivist
2. Cognitive perspective
3. Emotional perspective
4. Behavioral perspective
5. Contextual perspective

Benefits of e-learning versus traditional classroom setting

1. Reducing environmental impact
2. Quality education, made affordable
3. Convenience and flexibility to learners.

Communication technologies used in e-learning: Communication technologies are generally categorized as synchronized or a synchronous.

A synchronous activities use technology such as blogs, wikis and discussion bonds.

Synchronous: Involve the exchange of ideas and information with one or more participants during the same period of time. A face to face discussion is an example of synchronous communication.

Virtual classrooms and meetings can often use a mix of communication technologies.

Goals of e-learning : E-learning lesions are generally designed to guide students perform in specific tasks. Information based learning content communicates info to the students. Examples include content that distributes the history or facts related to a service, company or product. Information based content there is no specific skill to be learned. In performance based content to lesions build off a procedural skill in which the student is expected to increase proficiency.

Theories of mass communication (Message flow models)

1. Hypodermic needle model
2. One step flow model
3. Two step flow model
4. Multistep flow model
5. Theory of Mass media effects

6. Theory of individual differences
7. Psychodynamic model of persuasion process.
8. Theory of social categories.
9. Theory of social relationship
10. Socio-cultural model of persecution process.
11. Free-press theory
12. Social-responsibility theory
13. Development media theory
14. Democratic participant media theory.

Psychodynamic model: Explains about the relationship between internal processes and manifest over behavior of persuader.

Theory of social category:- Emphasis that people who have a number of similar characteristics are likely to have similar orientation and behavior resulting in fairly uniform response to mass media.

Information Kiosk

Features

1. Touch screen
2. Smarter buying and selling divisions
3. Increases farmer awareness
4. Printing on demand options for information kiosks
5. Self service kiosks reduce head count by lowering employee head count.
6. Reduces waiting in line by farmers.
7. Virtual sale assistance increases sale.

Phases/steps

1. Designing
2. Installing
3. Evaluating

Location: KVK/Veterinary centre/

Purpose: Easy access to necessary information

Goal: To strengthen the agriculture and irrigation sector for increase in agriculture production.

Facilities: 1. IT, 2. Training

Information available at information Kiosk/Centre

1. Project related activity
2. Printed material
3. Irrigation networks
4. Rule, regulation and agreement
5. Internet connection
6. Data about the availability of water at barrage.
7. Information regarding laser land leveling
8. A website of their own
9. Information about

MIS

Evolution 1990

1. Electronic data processing
2. Computer based information system
3. Decision support system
4. Artificial intelligence

Uses of MIS on different levels of management

1. Top management
2. Middle Management
3. Lower Management

Design

1. Assessing: Information needs for planning monitoring
2. Deciding: Deciding the levels of info
3. Ensuring: System flexibility and adaptability.

Guidelines for effective design of MIS

1. User participation
2. Cost evaluation
3. Alternative for in-house software development
4. Relevance and selectively of information
5. Pretesting of the system
6. Training the operators and users carefully

Classification of MIS

Mason and Swanson 1981 describe four categories of management information system:-

1. Database information system
2. Predictive
3. Decision Making
4. Decision Making

Roles of MIS

1. To plan the most effective allocation of resources.
2. To choose between alterative course of action.
3. To control day to day operations.

A typical functional structure of ATIC:

(A) Products	(B) Services	(C) Information
Seeds plant materials livestock species tools equipments value added products other agro products	Soil and water testing plant clinics very clinic testing and caliberationry small equipments	Exhibits audio, video, aids, form literature

Farmers and other end users

MAKS/AKIS

Rolling and Engel 1991: AKIS is a set of organization and or person and the link and interaction between them that are engaged in or managed such process on the anticipation, generation transformation, transmission, storage, retrieval, integration, diffusion, utilization of agricultural knowledge and information which potentially work systematically to support decision making, problem solving an innovation in agriculture.

Elements

1. Agriculture research
2. Agriculture education
3. Organization
4. Rural farming community
5. Mass media

Models of AKIS

1. The INTERPAKS MODEL SWANSON 1985
   - Policy Knowledge Transfer
   - An input transfer
2. Elliot's ATMS model: 1987

Functions/importance of MAKS/AKIS: - By Nigel 1980.

1. Identification of the problem/Knowledge needs of the procedures level.
2. Generation of innovation
3. Validation under farmer's condition
4. Operationalisation for utilization
5. Dissemination
6. Utilization
7. Evaluation of experience

Expert System: "Donald 1999 Expert system is sophisticated computer programmers' that manipulate knowledge to solve problems efficiently and effectively in a narrow problem area".

1. Planning
2. Designing
3. Selection
4. Diagnosis

5. Interpretation
6. Prediction
7. Decision technologies

Components

1. Knowledge base
2. Inference Engine
   i. Terminal graphic representation
   ii. Multiple character windows
   iii. Multiple graphic windows

Procedure: For development

1. Preparation of knowledge base.
2. Preparation of flow chart
3. Data entry into the program me module and computing of knowledge base where expert system are most effective.

1. DSS
2. Problem solving/diagnosis
3. Data analysis
4. Customer/product support
5. Background monitoring

Application in agricultural extension

1. Decision
2. Advice
3. Tech transfer
4. Information holding and maintaining

Demerits

1. Lack of common sense
2. Do not make creative responses
3. Domain expert do not always able to explain their logic and reasoning

4. Cannot adapt to changing environments unless knowledge base is changed.

Important examples of different expert systems.

- AGWATER
- CROPWAT
- CIIMWAT

Cyber extension tools

1. E-mail
2. Interactive expert system on crop pest and diseases.
3. Internet browsing for extension information.
4. Video - conferencing
5. Call centers, sitcom networks
6. Discussion groups and news groups

Broad classification of ICT initiative currently in place

1. Central government initiative
2. State initiative
3. NGO projects
4. Private and cooperative sector transaction
5. Pure information services.
6. Multiple rural services.
7. Knowledge networks and knowledge banks.

CE Extension tools

1. It is an established complementary and supplementary mechanism.
2. The launch of KCC has added a new dimension.
3. It can be safely concluded that Indian agriculture is on threshold of a new comm. mechanism.
4. Cyber extension will further improve the internal competitiveness and penetration of Indian products.

CE The Process

1. Develop state of the art infrastructure to connect key stack holders.
2. Creating ICT awareness in all development departments.
3. Creating information packaging mechanism.
4. Networking with E.G. initiative of concerned state district.
5. Create a nodal cell in each state to monitor the progress of CE
6. Identify a nodal coordinating committee for cyber extension in India.

Services provided benefits of CE

1. Providing interaction among SFF and other rural people.
2. Providing up to date news and information service.
3. A question and answer service.
4. Creation and maintenance of state database.
5. Providing the details of poverty alleviation schemes.
6. Providing status of various govt. programmers.
7. Hosting website by major institutions.
8. Launching online rural development and extension journals.
9. Providing internet access at district and block level.
10. Launching online journal of CE.
11. Providing internet access at district or block level.
12. Opening of cyber cafes
13. Providing maps that display different features such as population density crops planted etc.
14. Providing video clips to demonstrate complex procedure.
15. Providing user beneficiary feedback for the public sector scheme.

Development communication

The art and science of human communication linked to societies planned transformation from state of poverty to one of dynamic socio economic growth that maker for greater equity and larger unlading of individual potential.

Approaches

1. Information dissemination and education.
2. Behavioral change
3. Social marketing
4. Social mobilization
5. Media advocacy
6. Communication for social change
7. Participating development communication.

e-Choupal

e-Choupal is an initiative of ITC Limited, a large multi business conglomerate in India, to link directly with rural farmers via the Internet for procurement of agricultural and aquaculture products like soybeans, wheat, coffee, and prawns. e-Choupal was conceived to tackle the challenges posed by the unique features of Indian agriculture, characterized by fragmented farms, weak infrastructure and the involvement of numerous intermediaries. The programme involves the installation of computers with Internet access in rural areas of India to offer farmers up-to-date marketing and agricultural information.

Problems addressed

Traditionally, commodities were procured in *mandis* (major agricultural marketing centres in rural areas of India), where the middleman used to make most of the profit. These middlemen used unscientific and sometimes outright unfair means to judge the quality of the product to set the price. The difference in price between good quality and inferior quality was little, and therefore there was no incentive for the farmers to invest and produce good quality output. With e-Choupal, the farmers have a choice and the exploitative power of the middleman is neutralised.

Statement of a farmer who is a member of E-choupal: "Before ITC introduced us to e-Choupal, we were restricted to selling our produce in the local *mandi*. We had to go through middlemen and prices were low. ITC trained me to manage the Internet kiosk and I became the e-Choupal *Sanchalak* in my village. Today we are a community of e-farmers with access to daily prices of a variety of crops in India and abroad – this helps us to get the best price. We can also find out about many other important things - weather forecasts, the latest farming techniques, crop insurance, etc. e-Choupal has not only changed the quality of our lives, but our entire outlook.

Effects of e-Choupal

ITC Limited has now provided computers and Internet access in rural areas across several agricultural regions of the country, where the farmers can directly negotiate the sale of their produce with ITC Limited. This online access enables farmers to obtain information on mandi prices, and good farming practices, and to place orders for agricultural inputs like seeds and fertilizers. This helps farmers improve the quality of their products, and helps in obtaining a better price. Each ITC Limited kiosk having Internet access is run by a sanchalak — a trained farmer. The computer is housed in the sanchalak's house and is linked to the Internet via phone lines or by a VSAT connection. Each installation serves an average of 600 farmers in the surrounding ten villages within about a 5 km radius. The sanchalak bears some operating cost but in return earns a service fee for the e-transactions done through his e-Choupal. The warehouse hub is managed by the same traditional middle-men, now called samyojaks, but with no exploitative power due to the reorganisation. Indeed these middlemen make up for the lack of infrastructure and fulfill critical jobs like cash disbursement, quantity aggregation and transportantion.

Since the introduction of e-Choupal services, farmers have seen a rise in their income levels because of a rise in yields, improvement in quality of output, and a fall in transaction costs. Even small farmers have gained from the initiative. Customized and relevant knowledge is offered to the farmers despite heterogeneous cultures, climates and scales of production. Farmers can get real-time information despite their physical distance from the *mandis*. The system saves procurement costs for ITC Limited. The farmers do not pay for the information and knowledge they get from e-Choupals; the principle is to inform, empower and compete. At the same time ITC Limited has obtained benefits from the programme:

1. Elimination of non value added activities
2. Differentiated product through identity preserved supply chains
3. Value added products traceable to farm practices
4. E-market place for spot transactions and support services to futures exchange

There are presently 6,500 e-Choupals in operation. ITC Limited plans to scale up to 20,000 e-Choupals by 2012 covering 100,000 villages in 15 states, servicing 15 million farmers.

From the farmers hut

Few farmers like Abhishek Jainwho is a farmer of soya crop in Madhya Pradesh has got to say that e-choupal happens to be the best thing that has happened to him so far. Like any other farmer even he was constantly caught in the vicious web of the middlemen. With the help of ITC's initiative not only him but other farmers as well have been able to get a better price for their produce both in India as well as abroad.

Success that followed

Today this initiative of ITC has empowered the lives of people living in 10 states where 40000 villages have 6500 e-choupals and around 4 million farmers have been empowered. The initiative taken up by ITC to help the degrading reputation of a farmer has surely made a huge to difference to not only large but also medium and small scale farmers.

Milestones

Commencement of initiative: 2000

- States covered: 9
- Villages covered: 40,000
- e-Choupal installations: 6500
- Empowered e-farmers: 4 million

Through the e-Chou pal initiative, ITC aims to confer the power of expert knowledge on even the smallest individual farmer. Thus enhancing his competitiveness in the global market.

The immense potential of Indian agriculture is waiting to be unleashed. The endemic constraints that shackle this sector are well known - fragmented farms, weak infrastructure, numerous intermediaries, excessive dependence on the monsoon, variations between different agro-climatic zones, among many others. These pose their own challenges to improving productivity of land and quality of crops. The unfortunate result is inconsistent quality and uncompetitive prices, making it difficult for the farmer to sell his produce in the world market.

ITC's trail-blazing answer to these problems is the e-Choupal initiative; the single-largest information technology-based intervention by a corporate entity in rural India. Transforming the Indian farmer into a progressive knowledge-seeking netizen. Enriching the farmer with knowledge; elevating him to a new order of empowerment.

e-Choupal delivers real-time information and customized knowledge to improve the farmer's decision-making ability, thereby better aligning farm output to market demands; securing better quality, productivity and improved price discovery. The model helps aggregate demand in the nature of a virtual producers' co-operative, in the process facilitating access to higher quality farm inputs at lower costs for the farmer. The e-Choupal initiative also creates a direct marketing channel, eliminating wasteful intermediation and multiple handling, thus reducing transaction costs and making logistics efficient. The e-Choupal project is already benefiting over 3.5 million farmers. By 2012, the e-Choupal network will cover over 100,000 villages, representing 1/6th of rural India, and create more than 10 million e-farmers.

A digital transformation

ITC began the silent e-revolution of rural India with soya growers in the villages of Madhya Pradesh. For the first time, the stereotype image of the farmer on his bullock cart made way for the e-farmer, browsing the e-Choupal website. Farmers now log on to the site through Internet kiosks in their villages to order high quality agri-inputs, get information on best farming practices, prevailing market prices for their crops at home and abroad and the weather forecast - all in the local language. In the very first full season of e-Choupal operations in Madhya Pradesh, soya farmers sold nearly 50,000 tons of their produce through the e-Choupal Internet platform, which has more than doubled since then. The result marks the beginning of a transparent and cost-effective marketing channel. Bringing prosperity to the farmers' doorstep.

Linking farmers to remunerative markets

Farmers grow wheat across several agro-climatic zones, producing grains of varying grades. Though these grades had the potential to meet diverse consumer preferences, the benefit never trickled down to the farmers, because all varieties were aggregated as one average quality in the mandis. Enter ITC's e-Choupal intervention. The e-Choupal site is now helping the farmers discover the best price for their quality at the village itself. The site also provides farmers with specialized knowledge for customizing their produce to the right consumer segments. The new storage and handling system preserves the identity of different varieties right through the 'farm-gate to dinner-plate' supply chain. Encouraging the farmers to raise their quality standards and attract higher prices.

Managing risks through technology

The and ifs in the aqua farmers' life posed daunting odds. They were haunted by the nightmare of contaminated soil, wrong levels of salinity in the water or the killer White Spot virus, any of which could wipe out an entire shrimp crop, until the e-Choupal site provided them the support and the know-how to cope with and manage such risks. Information equips farmers with comprehensive know-how to keep abreast of food safety norms to compete in the international market. Information includes parameters for antibiotic usage, hygienic washing, sanitized dressing and air-tight packing. All these factors help to neutralize the risks involved in aqua farming. Making it economically much more attractive, benefiting hundreds of aqua farmers.

A dependable knowledge partner

Coffee planters in India have for years been tossed between the highs and lows of the international coffee market. The information needed to manage risks in the volatile global coffee market, price updates and prevalent trends in coffee trading were just not available to them. Launch of e-Choupal.com has equipped India's coffee planters with appropriate knowledge base and risk management tools. The site arms them with the latest prices posted on commodity exchanges like CSCE in New York and LIFFE in London. Planters have access to technical analysis by experts to help them comprehend trends, trading ranges and chart patterns in simple language. 'Parity Chart' and the 'Calculator' on the site convert the coffee prices quoted in international auctions into raw coffee equivalent for the benefit of the small growers in India. Traders net, a special link on the site, brings together a large number of coffee planters, traders and roasters, creating a virtual market for transparent price discovery. ITC empowers Indian coffee growers with expert knowledge in logistics and risk management, thereby enabling them to face global competition.

"A quiet digital revolution is reshaping the lives of farmers in remote Indian villages. In these villages, farmers grow soybeans, wheat and coffee in small plots of land, as they have for thousands of years. A typical village has no reliable electricity and has antiquated telephone lines. The farmers are largely illiterate and have never seen a computer. But farmers in these villages are conducting e-business through an initiative called e-Choupal, created by ITC, one of India's largest consumer product and agribusiness companies."

❑❑❑

Chapter 7

Farm Radio

Radio has been one of the oldest media access points for information seekers. With informationrevolution in rural India through radio has witnessed green revolution. Farmers were very keen totune the radio for forecast of weather, market prices and pests and diseases. But, with the limited timeslot and dominating function of entertainment through music, agricultural issues addressed remainedmore general and applicable mostly for irrigated belts. AIR has experimented many programmes forrural people. One such is Radio rural forums with a theme "listen, discuss and act". Forum of 20persons meet once or twice a week. First discuss issues of cultivation aspects for twenty minutes andin remaining time they get answers for their questions. Questions raised in the programme were answered in last ten minutes of the next progrmme. The success it received was beyond expectations. Encouraged with many such forums i.e., 7,500 forums were started in the country till 1960s. Later itlost its identity as transistors became cheap, a shift from group to individual listening took place. This lead to the operational problems. In early 1970s some stations in south started "Farm School on AIR"which consisted of a complete course on subject of relevance to farmers in different episodes. Withthe success experienced at Bangalore AIR station in 1978, Farm School on AIR is being run at different stations.Government of India established Farm and Home units in AIR in 1965 to communicatespecifically on agriculture and related technical information. Now it is regular feature/ unit in all AIRstations. Programmes of F&H unit include talks, discussions, and interviews with successful farmers, school on AIR and agricultural bulletins, etc. Four different programmes in morning at 6.35-6.45 am, afternoon 1.20-1.30 pm and in evening at 6.50-7.00 pm and 7.15 to 7.45 pm with local names specific to each state. For example it is Polam Panulu, Padipanta, Vyavasaya Suchanalu and Illu-Vakili, etc. in AP. In its effort to make the programmes more rural and farmer oriented and location specific, AIR has initiated many efforts mostly in strengthening its coverage and network throughout the country, but air time for agricultural/rural sector remained same, this led to the initiation to address onlylocation specific, agricultural and rural livelihoods through community participation. In India, the campaign to legitimise community radio began in the mid 1990s, soon after the Supreme Court ofIndia ruled in judgement of 1995 that "airwaves are public property". This came as an inspiration to groups across the country, but to begin with, only educational (campus) radio stations were allowed,under somewhat stringent conditions. Anna FM is India's first campus 'community' radio, launched on 1 Feb 2004. On 16 November 2006, the government of India notified a new Community Radio Policy which permits.

NGOs and other civil society organizations to own and operate community radiostations. In December 2002, the Government of India announced a policy for granting communityradio licenses to well establish educational and management institutes. In this direction few NGOs have initiated their own community radio centre. Frequency modulation technology has also broughtrevolution in localization of broadcasting and operation of large number of stations. However, thishas to be supplemented by a strong community radio movement to ensure greater involvement oflocal communities in the development process. Despite rapid technological changes in telecommunications in the last few decades, radio broadcasting remains the cheapest mode of information dissemination, equally catering to the needs of the rich and poor, rural and urban masses and reaching the remotest parts of the country. Needlessto say, it plays a vital role in the country's socio-economic and cultural development. Among thevarious modes of radio broadcasting, community radio especially has an important role to play. Due to its focus on local concerns and aspirations and the interactive nature of its programming, communityradio can be a powerful medium for education and development. The experience of a number of developing countries in using community radio for such purposes has clearly demonstrated its tremendous potential for strengthening development support for farming community. This module will certainly help you in understanding the concept of community, radio, different experiences in the country and enable to motivate communities for having community radio. You will be able to assist the communities in identifying needs, selection of topics, programme production and feed back assessment in community radio programmes.

Farm Radio-Meaning, Importance

Types of Radio

Radio stations can be categorised into three types:

Public radio : This is radio owned and operated by the government. Imain purpose is normally to servethe general public in the way that the government thinks is best. Private radio this is radio owned and operated by individuals or companies. The main purpose of this kind of station is to make money. *Community radio:* This is radio owned and operated by a community or members of a community. It can be supported by the state, individuals or corporates or even international bodies. The central purpose for this radio is to offer the people a voice and help develop the community. A radio station is recognised as 'community radio' when the station is owned by a non-profit group or by a co-operative whose members are the listeners themselves. Community radio is a broadcasting organisation established to provide communication support for the social, economic and cultural development of a community within a geographical location and owned and operated by the community on a non-profit basis.

Concept of Farm & Farm Radio

Origin of the word 'Community' lies in having 'something in common', something that can refer to geographical concepts like the neighbourhood, village, town, etc. but can also refer to other social determinants such as ethnicity, religion, gender, etc. Thus two forms of community – geographical communities and communities of interest - can be identified. Virtual communities, which display many of the characteristics of geographic or interest communities but are based in Cyberspace. When used in common parlance, 'community' means a body of people with some common interests. It is difficult to give an exact definition of community radio, since there are many differentmodels of community broadcasting. Community radio is characterised by access, public participationin production and decision-making, and by a non-profit listener-friendly economics. The management of the station is in the hands of those who use and listen to it. The workings of such stations are noteasy, but the structure does allow the audience/user to participate in the whole operation. Farm radio by itself may not be an end or only an independent means for achieving social development butit must be coordinated with ground level initiatives and has to be perceived as an additional arm for social communications and capacity-building programmes. It should be seen as a part of a broader struggle for access to communication media a struggle not only for freedom of communication, but particularly for the right to communicate, which has gained support in the past 25 years. Farm Radio essentially means a wireless broadcasting installation that is owned andoperated by the community that forms its audience. It commonly implies a low power and inexpensive setup that confines its signal to the community of people clustered nearby.

Farm Radio - Importance in Rural community

Farm radios across the world provide a range of diverse models. These are inevitable given that a Farm radio takes into account the social and political environment within which it operates. Consequently while the African experience may have seen a strong tradition of rural radio,Asia has experienced a mix of state public radio, quasi autonomous and independent Farm radio models. The Free Radio movement in Europe was a response to a centralized state owned broadcast system. At its height there were thousands of unlicensed private radio stations, but by the1980s this had dwindled to a little more than a handful. North America (notably Canada) and Australia have active community radio movements with a strong presence of co-operative radio. Despite differences in the models, there is a single cross–cutting factor: Farm radio as a critical voice incommunity development. It is called popular or educational radio in Latin America, rural or localradio in Africa, public radio in Australia and free or associative radio in Europe. All these describe the same phenomenon i.e., give a voice to those without voices, to marginalized groups and to rural communities. The National Program of AIR, though, was in Hindi and English, which, the policy framers.

Eventually realized, may not have been understood by a great number of people across the country, given the number of languages that are spoken. This led to the creation of Local Radio Stations (LRS), he third tier of broadcasting, set up as community stations providing utility service to a small area. The first LRS came up in Nagercoil, in the state of Tamil Nadu, a model that was replicated in various parts of the country, particularly small towns and rural areas. These LRS were meant to produce all their software locally, preferably in the local dialects. By 2003, a total of 78 LRS were set up, covering about 13.5% of the country's area and population. The LRS were a refreshing change from the some whatself important national and regional broadcasts. It was believed to be a mechanism for enabling the participation and the involvement of the local community, makibroadcasting relevant to theirlives. Though the LRS was a brilliant idea most of these Local Radio Stations are locked up due to lack of staff and other problems. Local radio or farm radio is a more recent concept of broadcasting in India. They are Flexible and spontaneous enough to enable the farm radio centres as mouth pieces of local community. Radio being limited in its time slot for agricultural programmes has restricted reach toonly assured irrigated belt. Considering the location specific needs of farming community and rain fedgeography the Government of India has opened up a window of community radio. Still the concept of community radio is not fully operationalised. Four modules of community radio were being steeredby some of NGOs viz., (1) Kutch Mahila Vikas Sangham (KMVs) in Kutc, Rajasthan; (2) Deccan Development Society (DDS), Medak, A.P., (3) VOICES in Boodikote, Karnataka and (4) Altematives Farm Radio-Meaning, Importance andfor India Development (AID) in Daltonganj, Jarkhand. These organizations have creatively used theinfrastructure of AIR or narrow casting to address location specific needs of community. The power of community radio lies in its participatory nature as both its content and technology are people oriented. It is an affordable means of communication, where people themselves raise issues and identify their own priorities. This has demonstrated a huge potential to fulfill the information and entertainment needs of the community. Here are some of the experiences of community radiomodels emphasizing its importance.

- Community radio is a catalyst for social development and is prevents to most obvious and some times the only choice to respond to the information and cultural needs of illiterate, marginalized, remote and under served communities. It is closest to oral culture and found to have huge potential for communication.
- Radio is an enormously supportive medium, particularly in context of poor women in their struggle for autonomous markets and legitimate space for discussing issues that main streammedia may not be interested.
- The centrality of community participation of content is a cross cutting non-negotiable in all the initiatives.
- This Community radio can also give traditional folk media a new lease of life by involving traditional folk forms and creating new sustainable hybrid forms of expression and communication.
- Community radio can become right platform to surface women's rights in socio-political situations.
- It can be effective and successful means of maintaining community cohesion during disasters and in times of emergency.
- Listeners feel greater sense of identification with this programme and clearly say that this isdistinct from AIR.

Expectations of Community Radio Stations

The government expects a number of benefits to flow from the successful operation of community radio stations across the country. They are

- Enhancing participation of people in the development process
- Capacity-building, especially in rural areas, through education
- Providing opportunities to people to upgrade their skills and enhance their creative talents
- Preserving and promoting traditional wisdom, knowledge and skills, thereby helping to Promote and project the local languages, arts, crafts, culture and traditions.

- Bringing within easy reach of the rural population, topical information in areas of agriculture, Social welfare, education, health and environment.
- Creating rural networks for cottage and village industries.
- Strengthening Panchayati Raj Institutions.

The experience so far has demonstrated that if the above objectives are to be realized, the government's policy must aim at a rapid proliferation of community radio stations across the country. As such, a fresh government policy exercise is underway for a framework for community radio to fulfills objectives focusing on the following areas:

- Liberalization of eligibility norms to involve reputed civil society organizations
- Analysis of the procedures for clearance for expediting issue of licences
- Capacity-building in the concerned organizations to enable them to design and formulate relevant programmes for community radio
- Building an interface with local development/Panchayati Raj organizations for mutual Benefit
- Ensuring the financial viability of such stations
- Network planning for community radio stations to ensure availability of frequencies for large numbers of such stations.

Profile of rural Community Radio

His profile of a community radio is determined by the community it serves and by the community that owns the radio station. Divergent approaches and formats in urban and rural contexts are, community radio's inclusive character. In many ways, this enables it to straddle the urban-rural divide and strengthen development based on integration, and not displacement. Urban community radio has to cater to a complex set of divergent activities: Community Radio-Meaning, Importance and needs to combine information with entertainment, develop innovative ideas in radio presentation and encourage continuous evaluation of community response through its programme structure. Rural communities, in contrast, are usually culture-specific, responding to needs, demands and even cultural traditions that often remain unnoticed by outsiders. The print media does not reach them and poor literacy, coupled with lack of adequate developmental infrastructure, makes it difficult for them to avail sophisticated communications technology. In many of these societies, the oral tradition of communication is still the most accepted and trusted mode of information dissemination and their spoken language a locally evolved dialect unique to the community. Community listening of radio ismore popular and community.

Centres like anganwadis and issue-based clubs characterise community participation. Gradually members of these clubs become producers of specific programmes related to issues relevant to women, children, people with disabilities, etc. Programmes are also narrowcast using audio cassette technology to facilitate community listening.

Check your progress

1. Mention different types of Radio?
   (Please read section 3.2.1 and answer)
2. What do you mean by community radio?
   (Please read section 3.2.2 and answer)
3. In your view what constitutes rural community profile?
   (Please read section 3.2.5 and answer)
4. What are the benefits that community can enjoy with community radio initiated and owned?

By them?

Operationalisation of Community Radio

Existing platform for community participation

Radio is unique mass media which has the quality of conveying latest information quickly and promptly. As radio is audio media it is effective and serves only active listeners. It is most inexpensive medium in reaching large masses in shortest time. It is most effective to stimulate and motivate farmers to participate in mass communication extension activities through announcements, publicity and propaganda. Radio continues to enjoy an edge over in terms of its speed, flexibility and reach over television. Even with advent of most sophisticated information communication technology, radio still could adjust itself and keep its position in the competing mass media. In fact, radio is part of our daily life. The transistor audience and audience on wheels for instance are at present dominant in most parts of the world. Broadcasting in its significance, reaches wide range of audience and its impact constitutes the most powerful communication media in India. Radio broad casting started in India way back in early 1920s in Bombay, Madras and Calcutta. The Government took over in 1930 and started it as Indian Broadcasting Service. Later it was changedas All India Radio (AIR) in 1936 and became popular as Akashavani from 1957. At present AIR has 277 stations constituting 114 regional stations, 77 local radio stations, 14 relay centres, 3 exclusive Vividh Bharathi centres and 5 community radio centres with 149 MW and 128 FM transmitters with broadcast coverage of 91.37 per cent by area and 99.13 per cent by population. Radio broadcasting has three-tire system viz., national, regional and local. The local radio is the most recent concept of broadcasting in India. Today, AIR covers the almost entire population, thisnetwork has to be made use of by extension officers like you to reach the farming community. Presently it is acknowledged to be the major source of farm information in country. Many farmers and fisher men today tune to Akashavani to hear the forecasts on which their livelihoods depend. It became anaccepted emergency service during natural calamities and disasters. Programmes for rural audience are broadcast from all AIR stations an hour or more with 7% of total AIR broadcast time. This led to Community Radio-Meaning, Importance and the creation of Local Radio Stations, the third tier of broadcasting, set up as community stations providing utility service to a small area. The first LRS came up in Nagercoil, in the state of Tamil Nadu,a model that was replicated in various parts of the country, particularly small towns and rural areas. By 2003, a total of 78 LRS were set up, covering about 13.5% of the country's area and population. The LRS were a refreshing change from the somewhat self important national and regional broadcasts. It was believed to be a mechanism for enabling the participation and the involvement of the local community, making broadcasting relevant to their lives. Though a few inspiring partnerships have evolved between Local Radio Stations and voluntary groups, such as in Palamau in Jharkhand, where two NGOs joined hands to launch a weekly community radio program called 'Chala Ho Gaon Mein', the current status on LRS is that plans to further expand the network have been put on hold. AIR has experimented many programmes for rural people. One such is Radio rural forums with a theme "listen, discuss and act". Forum of 20 persons meet once or twice a week. First discussan issues of cultivation aspects for twenty minutes and in remaining time they get answers for their questions. Questions raised in the programme were answered in last ten minutes of the next progrmme. The success it received was beyond expectations. Encouraged with many such forums i.e., 7,500 forums were started in the country till 1960s. Later it lost its Identity as transistors became cheap, a shift from group to individual listening took place. This lead to the operational problems.

Ownership & technology issues

Ownership issues

Community radio is always projected as a low-cost, non-profit, easy to run venture, operated primarily by volunteers from the community and sustained by a minimal paid staff. However, it does require an initial investment to create the awareness, set up the broadcast infrastructure and a regular flow of finance to cover the operational costs. To ensure community ownership, sustained involvement of a wide range of interest groups (NGOs, funding bodies, government departments, international agencies) in a wide spectrum of activities (training, procurement of equipments, licenses, etc.) isneeded. The State had tried to bring radio closer to people by setting up a series of local low power radio stations intended to serve communities living within a very close geographical range but the limited success of the AIR's LRS has actually helped to emphasise the owneship issue for small radiostations for community empowerment and greater accountability in public affairs. The broadcast laws have neither encouraged community participation in programmes nor rejected the concept. Community radio ideally should have a broad-based ownership, accessible to any member of the community. Established with well-developed guidelines, the station must Ensure that no particular group within the community becomes a dominating component in its operations and programming.

Under the existing rules, apart from Prasar Bharati, the following can own a radio station:

- Educational institutions, including IGNOU can get license to run non-profit radio stations and involve the neighbouring communities in programming

- Commercial houses, with majority share owned by Indian companies, can buy radio frequencies, as and when government auctions are held, and run a purely commercial channel. All these organisations, public, private or educational have their own methodology of assessing the relevance of their radio channels and every channel understands that without direct participation

By target listeners, the channel tends to lose its audience sustainability. Therefore, emphasis on participatory programming is fast becoming the most accepted format for radio. In the cities, this change in the pattern of programming does help satisfy the issue of ownership for a large section of urban communities. However in the rural sector, the concerns are more localised, the spoken dialect more area specific and problems directly related to the social structure within the community. The question of ownership takes a different route in the rural sector. Here ownership is not merely a question of sustaining a programme. It is the reason why a programme is in place. It is concerned with issues like:

- Community understands of what owning a radio station means
- Why they feel that a radio station is important for them, and
- Confidence within the community members regarding sustainability.

Rural communities have shown that audio-programmes made to suit their aspirations and needs can be devised through capacity building exercises and reached within the community through community listening or cable-radio. These are useful means to explore whether ownership of radio is accepted by the community as a useful model for participatory development activities. The process of ownership can only start after the community gets interested in the radio programmes and feels the need to sustain it. Ownership invariably throws up issues of finance, getting financial support to start a projectand management of the finance for sustaining the programme. However, long-term dependence on outside support invariably reduces the community's sense of ownership. It is advisable to simultaneously look for funding opportunities from the very beginning, through community activities and support. This is very clear from the experience from Sri Lanka Kothmale Community Radio (KCR) project. Community Radio-Meaning, Importance and Kothmale

Community Radio (KCR) project was set up in 1989 by the government-run Sri Lankan Broadcasting Corporation (SLBC). The staff and volunteers were recruited from the sons and daughters of plantation workers and farmers in the surrounding areas. Following theinitial success of the radio station, it was able to get a grant of US $50,000 from UNESCO in 1998. They utilized the fund for creating an internet hub and their own website. When the UNESCO funding stopped in October 2001; KCR had to find alternative sources of money. The project authorities worked closely with hotels, in the nearby hill town of Nuwaraeliya, to attract foreign tourists to the region through publicity both on radio and on the net and sponsorship for their programmes in return. The idea came from a senior community member who was also a volunteer for the station. There are other instances of successful partnerships in KCR. For instance, a weekly programme on human rights is funded by the University of Colombo's Human Rights Centre.

Technology issues

The recent guidelines of the Ministry of Information & Broadcasting have allowed recognized universities, colleges and residential schools to set up 'community radio'. These stations can broadcast educational programmes, debates, and cultural programmes which are non-commercial in nature. The radio station can also be used by the institutions to develop the community in and around the campus area. The role of the implementing agency for setting up a radio station begins only after permission has been obtained from the Ministry of Information & Broadcasting and a broadcast frequency for the radio station allocated. Its' not merely a question of finance, nor of the technologybut essentially the motivating factor that determines for the community what they want from acommunity radio initiative and how far they are ready to support it. The technology, especially when conceived for a rural/marginalized section of society, has to limit itself within, what may be termed as, a 'satisfactory level of technical production' which involves a combination of simple recording equipment and highly sophisticated production techniques. Unlike programmes of mainstream radio, community radio programmes need not be technically slick. The essential technology for community radio is limited to the handling of a portable cassette recorder in the field, ability to use the microphone for studio recordings and developing skills for basic production techniques. The management of the station also contributes to the sustainability factor as does finance. The degree to which the station management can inspire the community to become responsible tothe station is an index for measuring its capacity to remain in operation. The issue of finance is critical, both to start the initiative and to keep it running.

The hardware - A radio station has basically three components:

- Studio for live broadcasts or recording programme for later broadcasts
- Transmitter, to convert the speech into a high frequency signal for transmission and
- Tower with antenna. An antenna mounted on a high pole or a tower is used to send signal in a large area. In the case of FM broadcasts, allowed for Community Radio Station (CRS), the signal travels in the line-of-sight. As such, a higher antenna has the advantage of higher overage. Handling and operation of the equipment on a daily basis, in the studio and at the tower maybe done by volunteers from school teachers or students or the community. Various groups can beformed, namely for hardware maintenance, content creation, OB coverage, dubbing and recording. They will be trained by the service provider/consultant as a part of the total solution. Once the radio Station is fully functional; the board/society needs to look into the following aspects of running astation:
- Rules for staff and volunteer management
- Rules on fundraising (this includes advertising)
- Maintain liaison with the local administration
- Policy for programme production
- Prepare annual financial statement
- Attend important meetings with anyone who supports the station
- Regular meetings with the programme staff to guide and encourage their work.3.3.3 The stake holders the principal stakeholder in any community radio initiative is the community itself. Members of the community should be allowed to decide what kind of programme would interest them, whether a radio programme enthuses them, can help them articulate their problems and aspirations. The community alone can decide the format and scope of the programmes, the kind of entertainment

that Community Radio-Meaning, Importance and will enrich their own cultural heritage and yet keep them in tune with the mainstream, and the community alone will throw up the talents who can sustain the radio programme day after day. Community radio is a process. It is not simply about producing radio programmes to put on air. It is About developing a community by using radio. It is important to clearly identify the goals in any community radio project from the very beginning. It helps to bring together people who can support the cause and expand the network of people who could help sustain the programmes. The idea of the community as the principal stakeholder needs to be generated not only to create enthusiasm among the participants and the audience but also to help the community feel directly responsible. The key players in a community radio project that involves broadcasting are: Community members - This is the audience and the people who are central to the project Providers of technology - These people provide the equipment that the project will need; train members of the community on how to use and maintain the equipment Government - There are two levels in the government that need to be kept in focus: a) The department which will grant permission for the programme either through AIR channels or through ownership of radio transmitters : b) The local administrative authoritiesProducers and staff - These are the people who know the crafts of programme production andwill help the community to generate ideas and create programmes. They will also help with inputs from outside sources that would add to the general information pool essential to sustain radio programmes Resource mobilisers - These are the people who provide the finance to support the initiative. These could be the State, corporate house or NGO but the ultimate sustainability must come from thecommunity. These groups are the stakeholders. They are called stakeholders because each stakeholder hassome interest in running the programme, and a 'stake' in the project. They are all partners in theproject and share the responsibility of making it successful. Unless the participant, the listener and the broadcaster are all stakeholders in a community radio programme, it is difficult to sustain the programmes.

Content planning, designing and training

Content planning

Content is the heart and soul of any broadcast: non-profit community broadcasts, commercial channels and even community listening sessions are successful only when the content is interestingly designed and includes useful information. These aspects of audio production in turn depend upon the community the broadcast desires to reach out. Content, when planned by communications experts, offers many variations in presentation and style. Produced professionally, drawing participants from awide talent pool, and dealing with interests will benefit the listener, such programmes are models of

Creativity in broadcasting. Community radio programmes, however, are designed differently:

- Style of presentation has to follow familiar forms of oral communication
- Talent pool must be primarily drawn from the community
- Issues must excite the listeners and
- Broadcast has to relate to the daily life of the people it targets.

This is only possible when the community is an integral part of the management of a radiostation, providing inputs into content planning, exchanging ideas regarding presentation and also participating in the programmes to reach out with experience and information, sharing the joys and sorrows of everyday life.

Designing

Designing programmes for a community radio is like developing an operation plan, withinthe standard radio formats like interview, talk, discussion, drama, skit or feature, the community expects to hear familiar presentation styles. This has to be done through the use of the local dialect (especially in the rural sectors) and by using programme formats that are simple to follow. The dividing line between guidance to programme production and imposition of ideas is so diffused that there is always the fear that a programme can be participatory and yet not satisfy the community. The goal ofevery programme must be to keep listeners interested and encourage loyalty to the station. In the rural areas, ideas often come from local oral traditions and folk performances. In the urban areas, trends in media across the world play an important role in determining how the listener will respond to a certain style of presentation. The creativity of the professional broadcaster is a key Community Radio-Meaning, Importance and factor for designing urban community radio programmes while understanding the listeners' reaction to a programme is central to the rural programme designs. Some of the standard formats for radio production suitable for community broadcast are:

- *Drama recordings* - Recording a play inside a studio. This requires a fairly big space and more than one microphone to cover the 'stage'.
- *Panel Discussions* - Here people sit around (this is the panel) and talk about a certain topic on air. One member acts as the anchor to ensure that all the panellists are given a chance to speak, guides the speakers so that they Do not stay away from the core issues, and ensures that at the end of the discussion there is some solution in sight.
- *Music Recordings* - Most community radio stations record music from their communities. This often means bringing a group into the studio, audio balancing and regular research into the musical tradition of the community.
- *Live Broadcasting* - This is broadcasting live from outside the studio. Live broadcast requires someone in the studio to monitor what is going on air and even help the presenters outside.

The equipment for such an exercise is more sophisticated and the participants must be wellequipped to operate live broadcasts.

Traditional folk media can be given new lease of life by involving traditional folk forms and creating new sustainable hybrid forms of expression and communication through formats.

Training

Capacity-building for communities, both in the rural as well as the urban sector, begins with learning to articulate the needs, aspirations and frustrations. An important area is training for equipment handling, like the variations in the use of microphone to record interviews, music or discussion and drama. A successful recording makes immense impact both on the speaker and the recordist. Learning the technique of radio production is another area. Yet another area of specialization is developing skills in equipment maintenance and minor repair works. By far, the most important aspect of capacitybuilding that happens through radio is the ability to interact purposefully within as well as beyond the community, the skills that develop while conducting a discussion, by learning the craft of transforming squabbles into debates, and through the realization that the radio presenter can earn a special place as a community representative capable of leadership. Building capacity means improving abilities and skills. This is important because the more skills and abilities the station has, the stronger it is and this increases its chances of sustainability.

Key issues & important considerations in operationalisation

Each community is uniquely different in its perception of issues that are central to the community's well-being and development. Discussion among different groups within the community, donor agencies and community activists can help identify the first set of issues for projecting throughthe radio programmes. For example- Health and hygiene issues. Information on rural development and livelihood schemes, low cost technologies of agriculture.

Financial requirements for equipping and running a station or Studio: Divide the total amount into initial investments, running costs, depreciation of equipment (for re-investment), fees and taxes, etc.

Running expense: Estimate the different budgets, e.g. renting and supply (energy, water, etc.), programme acquisition, programme production, personnel (staff salaries, freelancers' remuneration etc.), service and maintenance, general administration (travel expenses, telephone), etc.

Funds utilization: Ensure that the annual financial plan is screened by a professional business consultant or chartered accountant.

Income Generation: Generating income has to be a two-fold strategy from the beginning.

Initially the funds may be contributions from donor agencies supporting the cause and the mission statement of the radio programme, and also from members of the community who believe that the programmes will benefit the region. This requires continuous effort, innovative ideas of fund-raising involving the community.

1. Community Radio-Meaning, Importance and Operationalisation of Community Radio Course
2. Long-term dependence on outside support invariably reduces the community's sense of ownership of community radio-comment
3. For creativity in broadcasting Community radio programmes, they are designed differentlyhow?
4. What are the Key issues & important considerations in operationalisation?

Experiences of community radio in India & abroad

Important initiatives of community radio in India

There are, at present, four sustained initiatives where community participatory radio/audio programmes have been used to expand developmental activities in the region. Foremost among these is the Pastapur programme of the Deccan Development Society(DDS) which is also in the Forefront of activism for ownership of radio waves by communities. The initiative in Bhuj by Kutch Mahila Vikas.

Sangathan is an attempt by a NGO to extend grass-root developmental work through participatory radio using the medium-wave broadcast channel of AIR. AID-Bihar's initiative in the Palamau district of Jharkhand is a first of its kind: a NGO using air-time from a local AIR channel on commercial terms and giving the community a sense of ownership of the air-time. The multidimensional media initiatives by VOICES, in Budikote, Karnataka, use all the three forms of participatory programming - AIR channel, Post Graduate Diploma in Agricultural Extension Management (PGDAEM) Cable connection in homes in selected villages and group listening. These interventions are examples of how community radio, if granted ownership, could become an ideal tool for development in rural India.

Case-I: Namma Dhwani

Namma Dhwani, initiatied by VOICES has used community radio to address local information needs about government schemes, sericulture, and has even linked up with the school in the area to enable children to make their own programmes. A cable connection has been laid between the audio center & the school. Teachers have used the medium to make model lessons. They have partnered with the Resource

Centre of MYRADA, an NGO working in the Boodikote area to build the capacities of women's self-help groups and build a process that focuses the priorities of women in these groups.VOICES has its own production centre and narrow-cast recorded programmes through tape recorders. The VOICES project in Boodikote has received funding from UNESCO and is in the process of setting up a cable audio network to enable every household to receive programmes to overcome the limited reach associated with tape-recorders. VOICES, for instance, has used participatory processes to train women from self-help groups in station management and content production. It has set up a management committee for community mobilization and for assessment of the programmes. The training by NGOs spans not only technical dimensions of programme production but also includes attention to perspective building about gender, art and culture etc.

The Namma Dhwani project is almost completely run by women who are members of the self-help groups associated with MYRADA's Community Resource Centre. The Namma Dhwani project has explored linkages with many institutions to address sustainability. For training and capacity- building the project has been supported by the All India Radio; the local Panchayat has given some space for running the project; the project has obtained techno-managerial support from CECI.

Case II. Kutch Mahila Vikas Sangathan Initiative in Bhuj

The, UNDP funded Kutch Mahila Vikas Sanghatan (KMVS) experience has highlighted the potential of radio for creating and keeping alive intangible Communities in the overall socio-economic context of the erosion of communities defined by geographic boundaries. It uses the Kutchi language in its content, which mainstream All India Radio does not.

The Drishti Media Collective provided early technical support to the KMVS team. KMVS feelsthat their radio experience has revitalized debates on culture and identity, through carefully researched and packaged content and thus has helped foster a sense of self-hood. The programme also affirms and validates pluralism by (re)constructing through the radio, a concept of an over-arching 'Kutchi' identity.

The KMVS experiment, which use AIR as a platform, have successfully used the radio as a medium to address governance issues, focusing on issues such as accountability or lack of responsiveness on the part of the administration. The programme content, in case of KMVS reflects the aspirations of people associated with the organisation - the field workers and the village collectives.

For the UNDP supported KMVS, making the programme content entertaining and reflective of the needs and aspirations of the community has assumed central importance given the extent of their reach. Their explorations in packaging content chart a course that has seen a shift from a drama to a magazine format. This shift has coincided with an emphasis on moving towards direct community participation through segments like Lok Math (people's opinion), Parda Farsh (Expose) and Charcha (debate). The challenge has been to keep listenership alive in 950 villages in Kutch. Also, the team has been grappling with complex questions in transforming cultural-specific content to suit programming needs.

The KMVS team, however, feel that newer media can also give traditional art and culture a new lease of life, by engaging with traditional forms and creating new more sustainable, hybrid forms of expression and communication. For instance, Musafari is one of the key dimensions of their magazine, which resurrects Kutch history, art and culture and tries to reinterpret them in the contemporary context. Spaces have been created within these slots to feature dying art forms such as Vai singing.

KMVS also believes that their approach has been to constantly ask questions, even of traditional legends, and thus the medium has been a platform to argue with traditions as well as modernity. As impact today, in the case of KMVS, women leaders of sanghatans, who are part of the radio experiment feel that they have acquired legitimacy among their counterparts working on other development issues, such as watershed development. Many Communities have also articulated that they feel confident of being able to run radio stations on their own, having acquired the requisite skills through their involvement in NGO-supported processes. The radio has been used as a medium to reinforce ethnic identity and promote community cohesion during times of religious tension. During the Gujarat riots, KMVS called upon the people of Kutch through the radio to invoke the values of tolerance and plurality embedded in their Kutchi identity.

A preliminary review of KMVS indicates that people repose faith in the transformatory potential of the radio. Both reviews also testify a wide listener ship. Community autonomy in production of programmes and content has meant the flexibility to respond appropriately and in a timely manner to community crises. Post-earthquake programming by KMVS has been a platform for exchange of ideas.

The experience of the initiatives and reviews indicate that women who have volunteered to be part of community radio initiatives feel the process has empowered them considerably. However, socio-cultural barriers at the community and household levels affect women's listenership negatively, a fact that was borne out in the evaluation of and KMVS experiences. Also, for KMVS another challenge has been to enhance women's sense of ownership vis-a-vis the programme.

Other initiatives

Deccan Development Society (DDS) in Medak, Andhra Pradesh; The Deccan Development Society (DDS) has used radio for discussing its development strategies, ecological practices of people, nutritional superiority of local crops etc. The perspective that informs programme production by women in DDS is heavily influenced by the development priorities that have emerged through their association with DDS. Considerable focus is given to indigenous knowledge about agricultural practices. DDS has its own production centre and narrowcast recorded programmes through tape recorders.

Alternatives for India Development (AID) in Daltonganj, Jharkhand.

Community radio has also supported NGOs like AID in their mobilization and conscientisation efforts in extremely tough contexts such as the remote parts of Daltonganj in Jharkhand where access to basic rights and services has been an uphill task for people. The AID project is supported by the National Foundation for India (NFI). A review of the AID initiative indicates how the listeners feel a greater sense of identification with this programme and clearly articulate that this was distinct from All India Radio. However, socio-cultural barriers at the community and household levels affect women's listenership negatively, a fact that was borne out in the evaluation of the AID experiences. The AID experiment which use AIR as a platform, has successfully used the radio as a medium to address governance issues, focusing on issues such as accountability or lack of responsiveness on the part of the administration.

Important initiatives of community radio in world

Community radios across the world provide a range of diverse models. These are inevitable given that a community radio takes into account the social and political environment within which it operates. Consequently while the African experience may have seen a strong tradition of rural radio Asia has experienced a mix of State public radio , quasi autonomous and independent community radio models. The Free Radio movement in Europe was a response to a centralized State owned broadcast system. At its height there were thousands of unlicensed pirate radio stations, but by the 1980s this had dwindled to a little more than a handful. North America (notably Canada) and Australia have active community radio movements with a strong presence of co-operative radio. Despite differences in the models, there is a single cross –cutting factor: Community radio as a critical voice in community development. It is called popular or educational radio in Latin America, rural or local radio in Africa, public radioin Australia and free or associative radio in Europe. All these describe the same phenomenon, gaining voices and democratizing communication on community scale.

Cases

Case1. Radio Lumbini

Radio Lumbini which is just half an hour by car from the birth place of Lord Buddha calls itself the Voice of Peace. It went on air in 2000, the first community radio outside Kathmandu. It has a 200 watt transmitter and can be heard in Lumbini, Butwal and Birgunj. The station has a full time staff of 19 and broadcasts about fifteen hours a day. The station manager is lawyer turned broadcaster Bharat Bhoosal. When asked if he still practises law at the bar, he replies that he hardly has time to go home.

Nepal has a very strong cooperative movement. Radio Lumbini runs as a cooperative, with 96 shareholding members. At the moment an investment of Rs 20,000 is needed to obtain membership, which can be made in instalments. 3000 rupees is for shares and the remaining 17,000 rupees is a non-refundable deposit. Initially the membership fee was as low as Rs 1000 and the increase has been gradual. Only about 10 percent of the members are women. There is an executive body elected for three years which meets every month to oversee the functioning of the station, look into its programme policy, take cognizance of any faults, and sort out differences with the local administration or political groups. There are also technical subcommittees that ensure broadcast quality. Day-to-day activities are carried out by the co-operative which is also responsible for the overall management and ensure community ownership. The co-operative model enables easy and quick decision-making as there is a strong sense of ownership and identity. The station is self-reliant as far as running costs are concerned, it accepts help from donors and sponsors for improving programmes and upgrading equipment. In the event of a crisis the co-operative has the right to float shares.

People from all political parties, castes and walks of life have obtained membership. However, the majority are farmers. In theory anyone in the area could become a member of the radio cooperative, but as Bharat says: "Before we entertain new members we check their background and make sure they are really interested in radio. If we conclude otherwise we discourage such persons."

Durga Pandey, a member of the board of management makes a strong case for the radio station: There is a certain kind of knowledge you need when you live in Nepal and Radio Nepal programmes do not provide that knowledge. "Station manger Bharat Bhoosal tells us how the radio station participated in a programme to revive the lift irrigation system in the neighbourhood". We joined hands with the local authorities of 17 village development councils. The programmes were presented in the Bhojpuri and Nepali languages. For six years the farmers had not paid their dues. The programmes highlighted the need to pay up. We went to the extent of warning the farmers that if they did not pay up without a valid reason we would broadcast the names of the defaulters. The majority paid up and the system was restored to a great extent.

As the radio station is situated close to India, it gets some advertisements from across the border. However, they charge an extra 10 percent for advertisements promoting Indian goods. The radio station has to pay a tax of 4% on revenue earned from advertising. In a few months, a commercial radio station will start broadcasting from the nearby town of Butwal, so competition is definitely on the way.

The Maoist movement has resulted in the emergence of taboo topics. Prior to the proclamation of emergency, there had been very vibrant discussions on sensitive issues. But according to the Station Manager, they now keep away from discussing or being critical of the King, the Judiciary, the multiparty system, and the Maoists. So far, the local Maoists have not interfered in the programming. The entire town comes to a standstill when strikes are called, but the radio station is allowed to function uninterrupted.

Radio Lumbini has been operational for 2 years. Break-even came after one year. Monthly income - 120,000 rupees. Monthly expenditure - 90/100,000 rupees. The main problem is capital expenditure.

Strengths: a) Decision-making is easy. If the station faces a crisis, the cooperative can float shares. b) There is a strong sense of ownership and belonging.

Recommendations on approaches to sustainability

Recommendations of The Workshop on Community Radio in South Asia Exploring the Way Forward' 21-23 February Kathmandu on approaches to sustainability:

1. Pursue diverse funding:
   a. Sponsorship
   b. Social marketing
   c. Donations
   d. Advertising
   e. Co-productions and partnerships
2. Sell cassettes/CDs
3. Share programmes with other stations
4. Charge membership fees and set up friendship schemes
5. Compile programme material for publication.
6. Use listeners' contributions
7. Rely on volunteers
8. Build links with educational institutions

Check your progress

1. Mention important initiatives of community radio in India?
2. What are the three segments/formats followed in participatory programming by KMVS teams?
3. When did and where Radio Lumbini started and what are its strengths?
4. What approaches you advocate for sustainability of community radio?

Selection of topics for programme production through Radio

Identification of Information Needs of the Farmers

- For any extension programme or extension activity to be successful it should be need based and location specific. All the efforts of programme production, time and cost involved will be futile if farmer needs are not considered and topic will be irrelevant and untimely. So as an extension officer/ volunteer before deciding the topic for production must identify needs of farming community.
- Before starting any alternative community communication system, it may be useful to find answers to the following questions:
- Is there a specific demand for information, education and/or entertainment in a given area that can be covered by broadcasting?
- Is radio the best way to respond to this demand? What is its special role in comparison to and in connection with other media and/or the traditional forms of communication?
- Are there special social issues that need to be promoted through radio broadcasts involving the community?
- What are the target audience's communication needs?
- Is it necessary to put a whole new channel on the air, or would it be sufficient to supply programmes to existing stations available locally?
- How would the programmes fit into the listeners' daily routine? What do they do at what time of the day? What social contact do they have? What other media do they use?
- If the community is taken into confidence from the very beginning, everyone around would feel that they have a real stake in the station. The local people, when involved at every step of building a station become stakeholders to the project, as it promises participation. With introduction of concept of community radio to meet location specific technology and information needs of farmers/community, identification of their needs become essential and a regular feature. Farmers needs can be identified using many participatory tools/methods or approaches. They are as follows:

Secondary data reviews: Through reports, books, news items, feed back of different TOT activities can reveal the areas of importance to the farmers.

Observation: Direct, participant observation or indirect observation can be taken up. As an extension officer you will be organising many extension activities, during which you can observe and identify their needs, culture, traditions and make a note of it.

Semi-structured interviews: It is an informal guided interview session, only few questions are predetermined as the course of interview takes the new questions are asked in response to answers from those interviewed. It may be individual farmer, key informants, group interview, and community meeting. Generally these are conducted by team of experts of different agriculture and allied sectors.

Participatory diagramming: Participatory transects, seasonal calendar, Venn diagram, participatory mapping- resource and social map are effective tools for farmers needs/priorities identification.

Brainstorming: Exploration of ideas in an unrestricted basis individually by group of farmers in a given period of time. Then the exhaustive list is presented to group for prioritization.

Participatory group discussion: All the members of the community are involved and made to participate in the discussion, as a facilitator initiate the topic of discussion and let the members carry on while you make note of important needs expressed by members.

Analytical participatory tools: Ranking techniques, time line, time trend, scoring and stratification can be used for identification of needs.

Workshop: Local community members, experts and outsiders and other stakeholders in agriculture may be brought together to discuss on the existing situation and asked bring about their needs and strategies to address those through effective feed back mechanism also future information needs can be identified. For this particularly an intentional effort on the part of extension officer is essential immediately after the broadcast of the programme.

3 Factors effecting the selection of topics/type of format (Segment)

Any topic or subject may not be suitable in all situations. There are several factors that influence an extension officer/community radio volunteer in deciding topic for the programme. Some times as an expert or extension officer, you may be confused in selecting the topic. Hence, knowledge on factors influencing may be of great use in selecting right topic for right people in right time.

Target Audience

The target audiences are the community people, farmers who listen broadcast programmes.

Hence, depending on the type of target audience i.e., the community topic should be selected. For this you should answer questions like what are needs/problems/ farming situation type of crops grown, etc., are to be considered. For this information/data collected using participatory tools/methods while identifying information needs is more important. Age, level of literacy of the target people what are their expectations, their knowledge, the stage of adoption, etc., are to be addressed in the topic.

Objective/Purpose

For any programme one must answer the questions why for and what for the topic is selected.

Another important aspect is after listening the programme, what is that expected of the farmers/audience, i.e., whether you want them to get aware/convinced/or to adopt technology or to use information for decision making. Accordingly, the type of topic and format changes.

Type of Learning Experience and Complementary Experience

One must able to decide what type of learning experience is needed. Such that farmer gets right message. Ex: is it just giving information, thorough understanding of demonstration, announcement, sensitize about any social issue, etc. Similarly, is there any access to the listeners the other media like print media, extension contact, etc., or not. Definitely this influence the intensity of focus needed on the theme, accordingly topic/format (segment) can be changed.

The type of structure or format (segment)

It is basically dependent on resources, time and equipment available. Even the knowledge and skill gap of farmers also influence the type of format accordingly topic is also required to be changed.

Visualisation

Visualisation and creative skills enable extension officer to visualize the complete version of the total programme. Good visualization enables the farm broadcast programme to assess texture with rich and right audio messages.

Some time the availability of artists, experts, crop growth stage, or activity, etc., also influence the selection of topic.

Duration

Definitely duration or time slot given also influences the type of format and the subject area/ topic to be covered. Generally any programme it should not exceed 20 minutes as it is difficult to sustain the interest of audience beyond this time. If topic is too big, then it can be fragmented and broadcasted in different episodes. This should be considered while scripting stage.

Resources

Resources include, human resources i.e., availability of resource persons/experts/local artists and technical and competent staff. Important non-human resources include infrastructure, transport and budget and their availability determines the type of topic/content to be covered.

If any topic is chosen considering all these factors would definitely result in timely, relevant and location specific programme, which is ultimately the objective of our communication.

Programme Production

Important Elements of Programme Production

There are five important elements involved in programme production. They are: (1) audience/ listeners, (2) extension officer/ expert/volunteer, (3) subject-matter/content, (4) Broadcast programme officer and his team, (5) infrastructure, studio/equipment and other facilities.

Target group or farmers the audience of the farm broadcast constitute basic element in any programme production meant for them. As we have already discussed how their information needs, agro ecological and farming situation, socio-economic status and their personal profile influences subject matter area/content to be covered in the programme. Representative farmers from the community may be involved not only in information identification, but also during programme production i.e., planning, script, preparation and especially while evaluating and modifying the programme. They should participate in giving realistic feed back.

Extension officer/expert of the organization that intends to broadcast a programme for farmer's plays a pivotal role in programme production .He is central and play a coordinating role in bringing out a good programme. They should consider all other elements and financial resources while deciding, planning, preparing, recording, editing, broadcasting and in reviewing the feed back of the programme.

A meticulous systematic designing is required. This indicates how important your role in bringing out programme for farmers. Broadcast programme officer and the team of concerned AIR station or even the privately hired consultants form another important element in production. However meticulously the script is prepared, programme is designed, if it is not properly recorded by competent team and executed effectively by programme officer, and the programme will not give desired result. Competency and availability of technical people should also be considered.

Subject matter/content/message/theme to be communicated is the heart to the entire programme production and broadcast. It should be precise, simple, and direct and must be relevant to the clientele needs. If subject matter is too long, it is better to break it in to different modules. All the other elements enable/facilitate in deciding the topic/ content.

With regard to infrastructure, without which nothing can be done, because programme production requires basically broad cast facility and other advanced digitized equipment availability.

The access facility to farmers to listen the programme is also another important aspect as the purpose of programme is to communicate/educate or inform farmers.

Thus these 5 elements play a major role in farm radio programme production.

Assessment of Programme

Assessing the effectiveness of the programme and evaluating it as against predetermined objective or purpose for which it is meant is highly essential and important phase in the process of participatory community radio farm programme production. But, generally persons involved in production are not particular in taking up assessment and evaluation. Analysis of the data obtained in this regard provide proper base to future programmes, information needs that are not addressed and other lapses in the programme since farm broadcast programmes are meant for rural audience, who are basically illiterate and semiliterate, it is very difficult to get the assessment or evaluation through structured schedules, only interviews (semi structured, focused group type), participatory group approaches like ranking techniques, preferences, brainstorming, etc., can be applied for informal evaluation.

The structured schedules can be used or given to community or group leaders, grass root level workers or the bare foot managers and obtain the feed back. Even the opinion of the experts in the particular field working in different organizations can be obtained through programme review committee meetings. The following is the proforma which can be used for evaluating farm broadcast programme.

Rating of participatory community radio farm Broadcast programme

Title:

Duration:

A. Attracts audience: Good Fair Poor
   1. Opening: Captures attention
   2. Arouses interest
   3. Purpose clearly mentioned
   4. Easily understood

B. Holds interest
   1. Interesting to hear
   2. Practical facts; worth knowing
   3. Appropriate to audience
   4. Clear and in logical order
   5. Presentation simple with local illustrations

C. Personal performance. I:
   1. Informal, friendly nature
   2. Talks easily and well placed
   3. Clear pronunciation
   4. Confident

5. Sincere
6. Competent

C. Personal performance. II: Voice Skills
  a. Voice
  b. Speed
  c. Clarity
  d. Synchronisation
  e. Modulation

D. Production skills
  1. Transmission smooth
  2. Well synchronized
  3. Well sequenced
  4. Well recorded

E. Impact
  1. Holds intended audience
  2. Tells story in one idea
  3. Good conclusion
  4. Well summarized
  5. Selection of suitable format to the topic
  6. Information and entertaining

G. Background music to different sequences

H. Appeal for adoption

I. Duration Sufficient, Not sufficient

J. Coverage of content, fully covered covered to some extent Poor

Feed Back Mechanism

The present feed back mechanism involves seeking the suggestions, clarifications at the endof the programme from the audience and reviewing the programmes in review committees with representatives from concerned departments and farmer representatives. These are done on a regular basis. However, with introduction of community radio concept, this has been strengthened under the central scheme on Mass Media Support to Agricultural Extension. Audience Research Units (ARUs) for AIR and DD are established to collect the farmer's feedback. At state and district level monitoring committees are mandated to guide and monitor the programmes based on feed back on continuous basis.

- Evaluation of community radio projects provides valuable feedback on the production process, and offers opportunities for new and innovative programming. Evaluation research enables the producers and managers of community radio to:
- Verify that the intended listeners are receiving the programme at convenient times
- Ascertain whether the content is being transmitted in appropriate language and format
- Establish the overall effect of the programme in the lives of the listeners
- Make necessary changes in the management and functioning of the radio station or the audio production center.
- Understand the economics of the whole project, convince sponsors/donors of its efficacy, and seek further support.

Acquiring knowledge merely about the extent of the message reach through quantitative methods may not facilitate in addressing local problems instead adopting qualitative methods such as in depth interviewing, focus group discussions, and participant observation regarding farm radio programmes will definitely help not only in assessing the feed back but also participatory planning by the community in future.

Research on audiences of community media projects must necessarily use qualitative techniques as they are more participatory and treat audiences not as objects and passive receivers, but as subjects and active interpreters of media content. One need not, however, abandon quantitative approaches altogether. Depending on the scale of operations of the radio station, a sample survey could give the station a quick measure of the audience and their listening habits.

Written feedback (letters) received by the station from the listeners. This, though, cannot be a reliable measure in itself. Evaluation of community radio projects, whether conducted by the organization itself or by an external agency, must include

a. Examination of in-house procedures, processes, personnel and
b. Assessment of perceptions of external public, i.e. the audience.

When programmes are on air regularly, some people in the station can be made responsible for interacting with the listeners' clubs who become their most important local resource. Offering some sort of incentive to listeners' clubs and using them as extension of the radio programme has been a successful sustenance strategy for many community radio stations in different parts of the world. A well-planned and implemented evaluation study of a community radio project is certainly a worthwhile exercise. It will ensure better programming, efficient message delivery, and more effective service to the community. It will help to make the whole project more participatory and give the listening community a sense of ownership in the radio programme.However, as an extension officer playing central role in programme production you have to make follow up interactions with farmers during your visits, regarding the programmes immediately after broadcast. Even the evaluation/ assessment schedule as mentioned in previous section can also be used for getting feedback from semiliterate farmers, experts, peer group.

Summary

Farmers are very keen to tune the radio for forecast of weather, market prices and pests and diseases. But, with the limited time slot and dominating function of entertainment through music, agricultural issues addressed remained more general and applicable mostly for irrigated belts. Radio is unique mass media which has the quality of conveying latest information quickly and promptly. It is most effective to stimulate and motivate farmers to participate in mass communication extension activities through announcements, publicity and propaganda. Radio continues to enjoy an edge over in terms of its speed, flexibility and reach over television. Even with advent of most sophisticated information communication technology, radio still could adjust itself and keep its position in the competing mass media. In fact, radio is part of our daily life. Presently it is acknowledged to be the major source of farm information in country. Many farmers and fishermen today tune to Akashavani to hear the forecasts on which their livelihoods depend. It became an accepted emergency service during natural calamities and disasters. Local radio or community radio is a more recent concept of broadcasting in India. They are flexible and spontaneous enough to enable the community radio centres as mouth pieces of local community.

In December 2002, the Government of India announced a policy for granting community radio licenses to well established educational and management institutes. On 16 November 2006, the government of India notified a new Community Radio Policy which permits NGOs and other civil society organizations to own and operate community radio stations.

The power of community radio lies in its participatory nature as both its content and technology are people-oriented. It is an affordable means of communication, where people themselves raise issues and identifies their own priorities.

Community radio is a broadcasting organisation established to provide communication support for the social, economic and cultural development of a community within a geographical location and owned and operated by the community on a non-profit basis. Community radio is characterised by access, public participation in production and decision-making, and by a non-profit listener-friendly economics.

Community radios across the world provide a range of diverse models. Despite differences in the models, there is a single cross -cutting factor: Community radio as a critical voice in communit development. There are, at present, four sustained initiatives where community participatory radio/ audio programmes have been used to expand developmental activities in the region. Foremost among these is the Pastapur programme of the Deccan Development Society (DDS) which is also in the forefront of activism for ownership of radio waves by communities. The initiative in Bhuj by Kutch Mahila Vikas Sangathan is an attempt by a NGO to extend grass-root developmental work through participatory radio using the medium-wave broadcast channel of AIR. AID-Bihar's initiative in the Palamau district of Jharkhand is a first of its kind: a NGO using air-time from a local AIR channel on commercial terms and giving the community a sense of ownership of the air-time. The multidimensional media initiatives by VOICES, in Budikote, Karnataka, uses all the three forms of participatory programming - AIR channel, cable connection in homes in selected villages and group listening. These interventions are examples of how community radio, if granted ownership, could become an ideal tool for development in rural India. These organizations have creatively used the infrastructure of AIR or narrow casting to address location specific needs of community. To ensure community ownership, sustained involvement of a wide range of interest groups (NGOs, funding bodies, government departments, international agencies) in a wide spectrum of activities (training, procurement of equipments, licenses, etc.) is needed. However, long-term dependence on outside support invariably reduces the community's sense of ownership. It is advisable to simultaneously look for funding opportunities from the very beginning, through community activities and support. The experience of the initiatives and reviews indicate that women who have volunteered to be part of community radio initiatives feel the process has empowered them considerably. Nepal has a very strong cooperative movement. Radio Lumbini runs as a cooperative, with 96 shareholding members. There is an executive body elected for three years which meets every month. There are also technical subcommittees that ensure broadcast quality. Day-to-day activities are carried out by the co-operative which is also responsible for the overall management and ensure community ownership. The co-operative model enables easy and quick decision-making as there is a strong sense of ownership and identity.

Private radio: This is radio owned and operated by individuals or companies. The main purpose of this kind of station is to make money.

Radio: It reaches mass and diverse audiences. The specialization of radio stations by listener age, taste, and even gender permits more selectivity in reaching audience segments. Since placement and production costs are less for radio than for TV, radio is able to convey public messages in greater detail. Thus, radio is sometimes considered to be more efficient.

Stakeholders: Stakeholders are those who has some interest in running the programme, and a 'stake' in the project. They are all partners in the project and share the responsibility of making it successful.

IDS (2002) Recommendations for the development of community radio in South Asia Report.

❑❑❑

Chapter 8

Traditional Media in Agricultural Extension

Introduction

The hold of tradition is still stronger in our country, particularly in the field of agriculture; ageold beliefs and practices have survived to a surprising level. Traditional beliefs, attitudes, values and folkways materially influence the progress of rural and agricultural development. A National Committee in its report on communication media in India said, "From the point of view of its greatest appeal to the masses and its quality of touching the deepest emotions of the illiterate millions, the medium of songs and drama is matchless"(Vidyalankar 1964, p. 35). The great task of extension education is to impart a desirable controlling influence on others through the use of effective extension teaching methods. It is better to have an inventory of the locally available and familiar traditional media for possible utilization in the field of agricultural communication. A proper blending of the traditional media with the electronic media is a welcome step in this direction. Besides, occasions such as agricultural fairs, farmers' days, exhibitions, fertilizer campaigns, etc can be used by the development agencies as a forum to disseminate agricultural innovations through traditional media to the farming community.

Traditional media is the media of people but modern media is for the people. Traditional folk media, thought to be dying bread in many developing countries, are currently being given a new lease of life. Current interest is not only in efforts to preserve this indigenous art but also to promote them as effective communication media to sell development massages. John A. Lent Prof. of Communication, Tuple University, U.S. has found folk media as agent of changes. Modern media is imported one whereas folk forms are generated from people. Hence, utilization of traditional media has widened its impact, even across cultural regions where actually they have survived. Those who want to use these effective tools of communication should identify the folk channels as most popular and fashionable and understand their real life situation and utilize them to get maximum advantage.

Traditional Media Concept, Meaning, Scope and Importance

Traditional Media concept

Traditional media means the mediums through which the cultural traits passed from generation to generation. It is born and expressed in the idiom of people's culture and has always seemed to entertain, educate and propagate the existing ideas and attitudes. Keeping in view their intimacy with the people at the local levels, folk media channels prove to be powerful tools of communication in the rural society.

Concept of Folk

'Folk' means race of people.

'Folk Music' – Music passed on from generation to generation. Any music composed on this idiom is folk song. A folk song is a song handed down to people.

Traditional Folk Media

Traditional folk media have no grammar or literature but they are surviving through oral and functional sources. It provides channels for expressing social, cultural, regular, moral, religious and emotional needs of a particular society or community; hence, they are the media of people unlike the modern media that are for the people. Folk media channels are powerful tools of communication, which play a crucial role in the dissemination of information in the rural areas catalyzing socioeconomic development, say the authors. These folk forms of media have proved to be the agents of change in rural societies. They are passed from generation to generation by word of mouth or orally. Now they are dying in developing countries. Traditional folk media are mostly used during religious or festive occasions. They are very specific in nature. These low cost media besides being credible in rural areas have potential for persuasive communication and get instant feed back.

Scope and importance of Traditional Folk Media

The folk art forms satisfy the innate desire for self- expression and also satisfy man's need for moral instruction combined with entertainment as well as for dramatic and the lyrical. In contrast with the electronic media, it preserves and disseminates in a lively manner the traditions and culture of our forefathers. Indian folk forms have a generous mix of dialogue, dance, songs, clowning, etc. So it is a very important method for communicating agriculture technology to the farmers and for the benefit of rural development programmes.

Experiences of researchers on traditional folk media proved that there is a great scope for imaginative use of traditional reforms. There is a need to narrow the gap that exists between the source and the receiver. This gap can be bridged ably by a coordinating agency at regional level with intended messages in their existing folk forms without disturbing the important characters and structures of traditional folk media. The agency should be composed of traditional folk artists, scriptwriters possessing essential knowledge of target audience, the traditional media and competency to transform the intended messages. People who are involved in this should be aware of folkways, dialect, mores and taboos of our rural and tribal communities for desired results.

The traditional media has always been used for moral, religious, agricultural and also for rural development. Rarely has it been resorted to for pure entertainment. It played an important role in many cultures. Some forms are dying out for lack of support and the onslaught of the modern mass media or as a result of disruptions in the society. The traditional media is personal, familiar and more credible with which the majority of literate and illiterate people in the villages identify easily. However, these forms need to be used with understanding and sensitivity.

Folk entertainments can provide fresh and interesting programme material for the mass media making them, more acceptable to both the rural and urban audiences. The folk media can be usefully employed to educate the non-school masses. In sum, the role of the traditional media in a country like ours is of paramount importance in enriching our culture and tradition as also disseminating information and educating our rural folk.

Characteristics of folk media

The important characteristics of the traditional media are as under:

1. The accessibility of the traditional media is very wide
2. It involves more than one sense; the potential for cognitive gain retention is possible but restricted
3. Potential for effective change in the attitude of the people is high
4. There is no change in the skill of the people
5. The maintenance of particular message is not possible
6. Interest arousal capability is very high
7. Range of mode choice is narrow
8. Less operational abilities and skill; and
9. Personal cost is not relevant
10. Folk media never die but continue in spite of several obstacles or objectives in modern age
11. In the process of transmission, the contents of the media become less distinct and fail to sustain their content and original characters
12. As long as the contents and formats satisfy psychological and social needs of people, they are carried forward by people themselves
13. All types of such traditional media have identified their worthiness as communication agents with commonly accepted behavioural patterns of their respective culture
14. More the folk forms come down as expressive channels of messages more they are regarded worthy of their use in the society
15. In modern age their content will be changed or distorted regardless of their structural characteristics
16. The styles of folk media are intensively associated with rural masses.

Check your progress

1. Traditional folk media are close to the minds and hearts of rural people- Yes/No Comment
2. What do you mean by traditional media?
3. Mention any four important characteristics of traditional folk media

Types of Traditional Folk Media

State wise traditional folk forms

The forms of folk media are many. Un-official estimate of folk media existing in India runs around 6000 in number. With a little variation, all these forms can be included into various groups such as - puppetry, drama, Harikatha (Story of God), story-telling, folklore, tales, ballad singing, pad recital (visual aid), kavad (story box),folkdance, folksongs, street play, riddles, proverbs etc. However, a glance through the table below further illustrates state-wise forms of folk media.

Andhra Pradesh: Vasthi Natakam (Veedi bhagawatham), Yakshagan-Beval Natakam, Kuchipudi

Assam : Ankya Nat (Bahwanas), Kirtania Natak, Ojapalli

Bihar : Bidasia, Saraikala Chhau, Jat-Jatini, Bidapad (North-East) Ramkalia (Ram Leela in Cruda style-North Bihar)

Gujarat : Bhavani

Haryana : Song (Sangeetaka), Naogal (Mono performance).

Himachal Pradesh : Karivala, Bhagat, Ras, Jhanki, Narnatra, Haran or Harin (Mask dance).

Jammu & Kashmir : Bhand Pathar or Bhand Jashna (Festival Bhands), Vatal dhamali (Sword-play).

Karnataka: Yakshagan (North Karnataka), Sannata (Belgaum and Dharwad dist.), Doddate-Bavalata, Tala Maddle or prasang (Indoor form, does not use costume, Dasrata, Radhana (closely resembles the Tamasha of Maharashtra).

Kerala: Koodiyattam, Mudiattam, or Mudiyattu, Tyattu, Tharayattam, Chavittu, natakam (Christan rural drama) Chakiva Kooth (Thullal) Kathakalli.

Madya Pradesh : Maanch, Nacha

Maharashtra : Tamasha, Lalit, Bharud, Gondhal, Dashavataar

Orissa : Fala jatra, Daskamthai, Chhau Mayurbhanj, Mangal Ras, Sowang

Punjab : Nautanki, Naqual, Swang (East Punjab)

Tamil Nadu : Therkoothu, Veethi, Veethi Natakam, Bhagawat Mela Nataka, Kurvanji (Kuram or Kulluve (dance drama)

Uttar Pradesh : Ram leela, Ras leela, Nautanki, Bhagat, Sang-Swang or sangeetaka, Janki, Naqqal and Bhand (different from Bhand or Kashmir)

West Bengal : Jatra, Puruliya, Chhau, Pala, Cambhira, Kbigan

Goa, Daman & Diu : Dashavatar, Tiyatra.

Different Types of Traditional Folk Media

Puppetry

It is one of the important traditional media for communicating technology to the farmers in the villages. Puppetry is believed to be the oldest form of popular theatre in India. The term, sutradhar (string holder) was used for the director or the stage manager in the live, classical (Sanskrit) theatre of ancient India (AD 100-1000). Experts reason that this use of the terminology derived from the puppet theatre is an evidence that puppetry preceded the live theatre in India.

In ancient India, especially in the South, the puppeteers were respected as intellectuals. They were usually patronised by kings and wealthy families and enjoyed immense prestige. The art of puppetry was popular both as pure entertainment and as an educational medium.

Puppetry is especially suitable for villages. It is an inexpensive activity. It is an easily acquired art and even crudely made puppets when played with a lively sense of drama can hold an audience.

The puppet play can impart lessons on health, literacy, agriculture, home-making; education, employment, rural youth activities, recreation, etc. There are many types of puppets: String puppets (originated in Rajasthan), Glove puppets (source unknown), Rod puppets (Orissa and Tamil Naidu) and Shadow puppets (Kerala and Orissa).

Puppets can make an impact, if properly used with the active participation of the local people.

Local dialect should be used if at all a foreigner' wants to stage a performance in a village. Puppetry attracts all age-groups.

Puppetry vs films: The Union Bank of India and the Life Insurance Corporation has used the medium of puppetry with tremendous success in Uttar Pradesh, to arouse the interest of the rural folk in bank savings and LIC policies. A pilot study by the Indian Institute of Mass Communication, New Delhi, on the comparative impact of puppetry and documentary films in two villages near Delhi showed that the cheaper traditional medium could be as effective as film. People in both the villages responded more favourably to the puppet shows, than the films. The finding has led the Films Division and the Childrens' Film Society to go in for puppet films. Doordarshan too capitalizes the puppet form in its programmes for children.

Drama

This is also a good source of communication, entertainment and education for the villagers.

The effectiveness of folk drama varies from message to message. One-act plays depict the village life, its problems and solutions; can be written and prepared with simple themes. The messages communicated through the traditional media gain access to the mind through both ears and eyes and the mind produces a sense of richness of meaning in the individual. This creates a high degree of interest and makes learning more permanent.

The subjects for drama stories may be on poor/less yields and the role of demonstrations on improved practices, working and organisations of the rural development programmes, role of democratic decentralization through village leaders, youth. Mahila and Krishak mandals, how to handle village problems like child marriage, wastage of money through ceremonies, dowry, solving problems of education, health, sanitation, role of co-operatives and extension wing and how villagers can be benefited from it, etc. Drama can be arranged at a common place in the village where the maximum gathering can be held. The time of presentation should suit the convenience of almost all villagers.

People feel happy if seeing, hearing and action goes simultaneously during the performance.

Folk theatre forms: Tamasha

Tamasha is an extremely lively and robust form of folk theatre, going back to over 400 years.

In this form, some topic is selected and then a chorus of six to eight male singers cum performers and two or more female dancer-singers make the troupe. The more vigorous form of tamasha is called song troupes and is believed to be the genuine or real form. It has a cast of three to six female dancersingers (one of whom is the star artist) a tabla player and a harmonium and tundun player. The story part is the next element. The most important is the performance. The story of tamasha in the form of agricultural dialogue, song and dance is based on stories derived from myth and folklore. It is in this section that comments are made on contemporary and social problems. Government-sponsored tamasha troupes also educate the masses on family planning. Tamasha needs no elaborate act or costumes and place and time.

Nautanki

Nautanki is like other folk drama forms. It has a simple dramatic structure comprising smallunits led by a Ranga or Sutradhara- the narrator. Music is of prime importance in this folk drama. The main musical instruments used are kettle drum and dholak. As in opera, the dialogues are sung to popular folk melodies but now even to film tunes. The nautanki form is popular in the fields of agriculture and rural development. It can be in the form of Doha, Thumari, Dadra, Sher, Gazal and Quawwali which are very common among the rural people in many states. This dramatic form may easily be adopted in the social, agricultural and rural development programmes. The people can be educated through the medium to bring about desirable changes in their behaviour.

Keertana or Harikatha

This is a religious folk theatre. The Keertana or Harikatha is a kind of concentrated drama or monodrama in which a gifted actor enters a whole series of characters. Saint Narada is believed to have invented and practised the form with great success. It is believed to have spread from Maharashtra to Karnataka and Tamil Nadu about 150 years ago. It is a mainly associated with the Bhakti movements in religion and literature. It was used by saints like Kabir, Guru' Nanak, Farid and Tukaram to preach the faith as also to bring about social reform and political change. It is such a potent weapon in social education that Lokmanya Tilak is reported to have said if he was not a journalist; he would have been a keertankar.

Harikatha is also used by Central and State governments to educate the masses on family planning, developmental activities, democratic values and national integration with help of Keertankar or Kathakars. All India Radio and the Doordarshan too are using the keertan form for broadcasts aimed at industrial workers and rural audiences.

Street play

In this form, one narrates a story which is related to a particular theme. These plays are not like theatre but they attract a large number of people. They also have some music built in by way of folk songs adapted to the theme. The only material used in this form is a number of banners on which slogans are written. It is a non-costume, no-property no—script play. It is also known as Nukad play.

Street plays are increasingly used to spread messages of family planning, sanitation, adult education, etc and they create awareness against social evils.

Folk dance

During the harvesting season and festivals, folk dances are aroused in the villages as they are helpful in collecting the people. The youth present Bhangra during the harvesting period.

The folk dances reveal not only the individual talents of the people but the collective traditions of all the countries, the characteristics of the community and a love for rhythm. Folk dances are mainly grouped under three heads:.

i. Community folk dances held on the main religious festivals, social occasions, Kisan Melas and other occasions when special agricultural activities are carried out;

ii. Folk dances' are also preserved by hereditary, professional families and groups who perform at birth, betrothal and marriage ceremonies in the villages; and

iii. Tribal dances rooted as they are expressive of their magical philosophies of life.

Folk song

The villagers have a great fascination for their folk songs. We go closer to them if we participate with them and organize such functions during exhibitions, meeting, film shows, drama, etc. Songs connected with agriculture and rural development programmes and practices in local dialects can be composed and sung easily as they also provide entertainment. This is a good way of conveying information to the villagers. Song competitions can also be arranged.

Some audio cassettes produced by the Punjab Agricultural University are also based on some common songs entitled as Balliye Kanak Diye. It was well-received by the villagers as it contains information on various aspects of wheat cultivation, a major crop of North India, blended with rich folk music. The other cassette is Khiran Kapah Narme. In this, information about cotton is given in an attractive manner. Folk songs should have local time, tone and dialogue.

They are a good source of attracting the people and as such help extension programmes indirectly. They are part of cultural programme. The monotony of discussion of serious nature in a meeting can be broken by light songs. The song should be composed on the subjects which are to be communicated to the people. They should be in the form of a story with some moral. The tune of the song should be popular and local to which the villagers are accustomed to. It will always be preferred if the song has just one message to convey. A few words by way of explanation will be appreciated before singing is started. The song can be followed with a little more explanations of its theme.

Seasoned singers are liked by every body but new voices are also welcome.

Story-telling

Story-telling has been one of the best and most commonly use method of instruction in informal education, religious propagation, rural development, etc. Over the years, certain etiquettes have been developed that are associated with story-telling. Before the story teller begins the narration, he asks, are you ready to listen? The listeners respond with 'Yes' sound. This ensures proper attention when the story is narrated. The listeners often make some kind of sound at regular intervals to express attention. So important topics which help in proper development of the rural people can be touched through the stories.

Proverbs which predominate in oral civilization represent the essence of rural wisdom and knowledge. They are the sound symbols of culture and have survived for centuries; their use is quite frequent in oral culture. Proverbs are very creative in knowing the importance of information or idea as they also tell the meaning of the real message. Some proverbs on agriculture are given below:

On importance of farming

Raj ke wah, to dab ke kha

Bioscope

This is also a popular folk medium used for entertainment and for propagation of information on education, agriculture, etc. Bioscope consists of a box made of light wood. It contains quite a number of folding doors, each like a panel hinged on one side with another panel with colourful figures and mythological episodes and incarnations of gods and goddesses painted and filed as pictures.

The doors are unfolded evoking the curiosity of the audience as preceded with the narration of the story by the story teller. This is helpful in conveying the people about educational messages like vegetable preservation, mosquito control, cultivation of different crops, etc. This can be made interesting with musical narration. The narrator unfolds each door and conveys the message to the audience accordingly.

Munadi (Announcements)

This medium is very old. In this form, the drum is beaten and the attention of the audience is arrested and then the message is delivered. This medium is normally used to inform the people about some happenings or other important extension activities. This is mostly preferred by the rural; people as it is used for dissemination of information followed by drum beating eg for giving message about meeting, demonstration, field-day or any other activity. In these days in Punjab, announcements are given through-mikes or public address system from religious institutions; Chowkidars of the village also do this job.

Wall paintings

The use of this medium has possibly evolved from an age—old practice of drawing or writing on walls. They are silent unlike traditional theatre and if they use words, they make themselves meaningless also to the non-literate majority. Yet, they are near permanent. A speech or film comes to an end soon, but the wall paintings stay as long as the weather or the organisation concerned allows them to last. Perhaps, the greatest advantage of the medium is the power of the picture completed with its local touch. The images used have strong emotional association with the surroundings, a feat impossible for even a "moving" visual medium like television which must use general images to cater to the greatest number of viewers. Illiteracy, of course, is no bar for the wall paintings, if the message is conveyed by the picture alone and the words are only an adjunct. Indian experience in wall paintings communication by the government can still activate its potential in rural areas.

Traditional Folk Media Vs Modern Media

Distinguishing features between traditional folk media and modern mass media

Modern media Traditional media

1. Disseminate messages to the heterogeneous audience
2. Caters to the ethno-rural audience through functional role of folklore
3. Communication is through electronic and print media
4. Through vocal, verbal, musical and visual folk forms
5. Through organized sources of diversified nature through generation to generation from one mouth to others (Oral)
6. Most modern modes
7. It is shaped by organizations and institutions
8. It is shaped by cultural heritage and common ties of existence of people

9. High cost media
10. Low cost channels
11. Credibility is less
12. Credibility is more
13. Less effective in promoting the change 10. More effective in promoting change
14. Rich in variety and relished by all groups of people
15. Specific to some groups
16. Difficult to get feedback
17. Greater potential for persuasive communication. Face to face communication and feedback.
18. Any time the media can be used and specific to any occasion
19. Used during rituals, festivals, ceremonial functions
20. for literates (Usually)
21. Mostly meant for illiterates

Indigenous modes

Traditional Folk Media in Agricultural Development

Use of Traditional Folk Media in Modern Media

The integration of modern media and existing traditional media creates a process of interaction. Traditional folk media, through to be a dying art in many developing countries. But as per John Lent, Professor of Communication, USA, a new lease of life is given to this media. The Third WorldCountries are gradually giving serious thought for the use of traditional media to aid the development programmes. He further points out that it is India to lead in attempting to preserve the cultural identity of indigenous arts to bring a sense of functional relevancy to them and established a song and drama division as a branch of information and broadcasting ministry in 1954. This division is putting every effort to convert this ancient wealth of performing arts to developmental communication functions.

The foremost user of the traditional media in our country is the Directorate of Field Publicity and the Song and Drama Division of the Ministry of Information and Broadcasting, Govt. of India. The Regional Officers of these organisations work in collaboration with the field units and are also helped by the Block Development Officers, village level workers (Gram Sevaks and Sevikas) Agricultural Extension Officers and other local people.

The folk media programmes are scripted by "the artists themselves; the direction and treatment have to be approved by extension workers and local officials. They have proved successful in reaching out the rural and semi-urban areas. The various programmes include folk plays, poetry recitation, quawwalies, puppet shows, religious discourses like Harikatha, folk songs and folk dance.

However, the fact remains that the exposure of the rural population to social education and development programmes through the folk media is as insignificant as to the modern media.

Advantages and disadvantages of traditional folk media

Advantages of traditional folk media

- Most intimate and close to masses in all the regions in the country
- They are close to the heart and minds of people as their primary appeal is emotions but not intellect
- There is clarity in communication with their familiar format, content, local colloquial dialects
- As these forms are specific to different groups and regions there is no cross cultural communication hurdle
- Rapport is immediate and direct, the barriers to communication almost non-existent
- Easily available to their clientele
- Flexible to accommodate any new theme
- Generally all age groups enjoy these traditional folk forms, is it not ?
- They are low cost media compare to modern mass media
- The Indian traditional folk forms have a generous mix of dialogue, dance, song, clowning, moralizing and prayer
- Folk media unlike electronic media invite audience participation also
- Folk forms satisfy the innate need for self-expression for moral instruction through entertainment and for the dramatic and the lyrical

- These are the main vehicles for passing on traditions and culture from generation to generation.

Disadvantages of traditional folk media

Threat to traditional folk media is a regular with the development of film industry, radio, television and video, the Indian traditional media gradually disappearing.

1. Focus of attention of planners on modern media.
2. Scale of reach to masses is very less unlike as in case of technology based electronic media.
3. Changes in social system such as urbanization and literacy rate leading to more individual listening or viewing rather people are preferring community listening or viewing.
4. Considering the traditional media as out dated and irrelevant by the present generation.
5. Traditional folk media is functioning mostly as unorganized sector.
6. Even to revive traditional folk forms identification and selection of good artists is very difficult.
7. Lack of encouragement to talented artists in traditional media leading to decline of folk media.
8. Lack of proper understanding of traditional media and its nature.
9. Range of choice of forms is very low.

Blending of Traditional Folk Media with Modern Media

A wise strategy is one which bring about a mutual reinforcement of the advantages of both traditional and modern media. Both media are simply two different points on the same continuum.

To-day's innovation, is likely to be tomorrow's vintage medium (Best of typical medium). As per Shyam Parmar "the judicious use of mass media a reality. He further says gear the messages and give them different treatment according to audience having different norms in different settings. Shaping of communication contents should be negotiated through the prevailing systems of communication in different societies". Traditional media co-exists with rituals and ceremonial functions. Their audiences are regional and ethano-linguistic. They are persisting with flexibility to infuse any message for modernity. Though there are sporadic efforts made in this direction, a systematic and broad based efforts listed below are essential to make use of folk media in agriculture development.

No mass media can exist in vacuum. The multi-divisional cultural fabric of the Indian society as a whole is evidently deep set with certain values and norms. What is relevant to the Orissa tribe may not be relevant to the Todas of Nilgiri hills. The London expert group meeting on Inter Regional Seminar-cum-workshop on the Integrated Use of Folk and Mass Media recognized that "No communication strategy would be complete unless it included traditional media". For effective bridging of the gap which exists between new technology and its use by farmers, and dissemination of sociopolitical and scientific information a coalition of modern technology based mass media and traditional modes of communication is essential. In the total communication network, avoidance of folk media may keep the major part of the population per haps ignorant of many useful things.

However, mooting the idea of blending both traditional folk media and mass media is excellent and easy but process of integrating them without losing the individual identity is an up-hill task. A constructive and comprehensive role of all the personnel involved with mass media, traditional folk media and subject matter is desirable.

Don't you feel now, that traditional folk media forms once produced or developed can be multiplied or replicated through modern media like radio and TV at different places? A bit of focus and attention by all of us is needed to give a push to these traditional forms and use them as medium for dissemination of agricultural technology either by your own department or organization or in partnership with others.

Hence there is a need to use this traditional folk media with modern electronic media for more effective communication.

Sustaining Traditional Folk Media

The persistence of traditional folk media in spite of attack from technology based mass media is itself a strong evidence for their existence and survivability, like old habits, they die-hard. Though there are sporadic efforts to improve and make use of traditional folk media hitherto, a systematic and broad based efforts listed below are essential to make best use of folk media in rural developmen process.

1. A systematic search for preparing an inventory of folk media is essential which can be utilized in planning communication process.
2. Studies to identify the suitable folk media and flexibility to infuse scientific knowledge are to be encouraged.
3. The identified folk media with flexibility to infuse scientific knowledge are to be integrated with mass media and make case studies to study their utility to ruralities and improve upon further.
4. Conduct regional and national level seminar-cum-workshops to isolate and identify resourceful artists and their folk media.
5. Offer award and rewards to encourage resourceful folk artists of rural areas.
6. Provide incentives to the persons integrating mass media with folk media.
7. Most of the present mass media advertisements involving folk forms are mis-leading the common man in rural areas for their respective contents and quality of products which need to be screened thoroughly and controlled.

However, the entire process of making inventory, processing, isolating and integrating of folk media forms can be best undertaken in all nationwide rural development and agricultural programmes.

Traditional forms of entertainment have been used to further programmes of development.

Though their heritage has been in the narration (In classical myths and legends), it was discovered as far back as 1954 that they could be used to present developmental messages as well. The Song and Drama Division was set up as an arm of the Ministry of Information and Broadcasting with the responsibility of harmonizing the county's performing arts for the development of communication in 1954. The Song and Drama Division works in close cooperation with the field publicity offices which arrange various programmes. However, efforts have been made by other agencies such as the Indian Institute of Mass Communication, the Union Bank of India, Coromandal Fertilizers, etc. to reach the people through the traditional media. It is better to have an inventory of the locally available and familiar traditional media for possible utilization in the field of agricultural communication. A proper blending of the traditional media with the electronic media is a welcome step in this direction. Besides, occasions such as agricultural fairs, farmers' days, exhibitions, fertilizer campaigns, etc can be used by the development agencies as a forum to disseminate agricultural innovations through traditional media to the farming community. Indian folk forms have a generous mix of dialogue, dance, songs, clowning, etc. So it is a very important method for communicating agriculture technology to the farmers and for the benefit of rural development programmes. The traditional media should be an integral part of any rural development programme. Wherever possible, it should be integrated with the modern mass media; but in all cases, integration with the ongoing extension work is vital to sustain these media.

The poorest man has access to his culture, expressed either in story, poem, play, song, custom, rituals or a variety of other forms of characteristics of folk culture. So the traditional media plays a prominent role in rural development which needs some evaluation to assess its effectiveness for suggesting appropriate folk forms for various development messages. The traditional media has always been used for moral, religious, agricultural and also for rural development. Rarely has it been resorted to for pure entertainment. However, these forms need to be used with understanding and sensitivity.

Vulgarization can easily set in if they are left to the mercy of the urban elite. A deliberate distortion of the forms could easily alienate the masses. Meddling the folk forms which are alien to the idea of communication on contemporary topics must not be allowed. It must also be noted that folk forms are community and language specific and bear values and norms, which need to be taken into account while developing media strategies for development programmes.

The employment of traditional media for effectively conveying rural developmental messages in a language and style that will be comprehended and liked by the rural people leads to spectacular results. The feasibility of the traditional media in the development process is possible in agriculture, medical and health programmes including family planning, cleanliness, etc food and nutrition programmes, adult education, political awareness, social legislation, people's participation in rural development programmes, etc.

If the traditional media is infused with new themes, new purposes with financial and other back-up support, it can go a long way in preserving the invaluable treasures which are a source of pride to the people, especially the rural poor, who may not feel at home in a world of modern mass media. In sum, the role of the traditional media in a country like ours is of paramount importance in enriching our culture and tradition as also disseminating information and educating our rural folk.

Summary

A National Committee in its report on communication media in India said, "From the point of view of its greatest appeal to the masses and its quality of touching the deepest emotions of the illiterate millions, the medium of songs and drama is matchless". Traditional folk media, thought to be dying bread in many developing countries, are currently being given a new lease of life. Modern media is imported one whereas folk forms are generated from people. Hence, utilization of traditional media has widened its impact, even across cultural regions where actually they have survived.

Traditional media means the mediums through which the cultural traits passed from generation to generation. It is born and expressed in the idiom of people's culture and has always seemed to entertain, educate and propagate the existing ideas and attitudes. Keeping in view their intimacy with the people at the local levels, folk media channels prove to be powerful tools of communication in the rural society.

Traditional folk media have no grammar or literature but they are surviving through oral and functional sources. It provides channels for expressing social, cultural, regular, moral, religious and emotional needs of a particular society or community; hence, they are the media of people unlike the modern media that are for the people. These folk forms of media have proved to be the agents of change in rural societies. These low cost media besides being credible in rural areas have potential for persuasive communication and get instant feed back. Some forms are dying out for lack of support and the onslaught of the modern mass media or as a result of disruptions in the society. The traditional media is personal, familiar and more credible with which the majority of literate and illiterate people in the villages identify easily.

It involves more than one sense. It has potential for cognitive gain retention but it is restricted.All types of such traditional media have identified their worthiness as communication agents with commonly accepted behavioural patterns of their respective culture. Potential for effective change in the attitude of the people is high. The forms of folk media are many. Un-official estimate of folk media existing in India runs around 6000 in number. With a little variation, all these forms can be included into various groups such as - puppetry, drama, Harikatha (Story of God), story-telling, folklore, tales, ballad singing, pad recital (visual aid), kavad (story box),folkdance, folksongs, street play, riddles, proverbs etc. The integration of modern media and existing traditional media creates a process of interaction.

India should lead in attempting to preserve the cultural identity of indigenous arts to bring a sense of functional relevancy to them and established a song and drama division as a branch of information and broadcasting ministry in 1954. This division is putting every effort to convert this ancient wealth of performing arts to developmental communication functions.

A wise strategy is one which brings about a mutual reinforcement of the advantages of both traditional and modern media. Both media are simply two different points on the same continuum. To-day's innovation, is likely to be tomorrow's vintage medium (Best of typical medium)

Shaping of communication contents should be negotiated through the prevailing systems of communication in different societies. They are persisting with flexibility to infuse any message for modernity. Though there are sporadic efforts made in this direction, a systematic and broad based efforts listed below are essential to make use of folk media in agriculture development. No mass media can exist in vacuum. What is relevant to the Orissa tribe may not be relevant to the Todas of Nilgiri hills. The London expert group meeting on Inter Regional Seminar-cum-workshop on the Integrated Use of Folk and Mass Media recognized that "No communication strategy would be complete unless it included traditional media".

However, mooting the idea of blending both traditional folk media and mass media is excellent and easy but process of integrating them without losing the individual identity is an up-hill task. A constructive and comprehensive role of all the personnel involved with mass media, traditional folk media and subject matter is desirable. Though their heritage has been in the narration (in classical myths and legends), it was discovered as far back as 1954 that they could be used to present developmental messages as well. The traditional media should be an integral part of any ruraldevelopment programme. Wherever possible, it should be integrated with the modern mass media; but in all cases, integration with the ongoing extension work is vital to sustain these media. Rarely has it been resorted to for pure entertainment. However, these forms need to be used with understanding and sensitivity. It must also be noted that folk forms are community and language specific and bear values and norms, which need to be taken into account while developing media strategies for development programmes.

The feasibility of the traditional media in the development process is possible in agriculture, medical and health programmes including family planning, cleanliness, etc food and nutrition programmes, adult education, political awareness, social legislation, people's participation in rural development programmes. If the traditional media is infused with new themes, new purposes with financial and other back-up support, it can go a long way in preserving the invaluable treasures which are a source of pride to the people, especially the rural poor, who may not feel at home in a world of modern mass media. In sum, the role of the traditional media in a country like ours is of paramount importance in enriching our culture and tradition as also disseminating information and educating our rural folk.

For agriculture and allied activities folk media can be used to the fullest extent. Extension personnel can come out with a strategy which utilizes the mutual reinforcement of the advantages of both traditional and modern media. Traditional folk media play important role in influencing the behaviour of rural people including farming community. Don't you feel now, that traditional folk media forms once produced or developed can be multiplied or replicated through modern media like radio and TV at different places? A bit of focus and attention by all of us is needed to give a push to these traditional forms and use them as medium for dissemination of agricultural technology either by your own department or organization or in partnership with others.

❑❑❑

Chapter 9

Model Questions for UGC/ICAR-NET/JRF/ARS

1. The element that distorts or disrupts a message being sent from a Sender to a receiver is:

   (a) Cognitive (b) Medium
   (c) Noise (d) Resonance

2. The process of controlling the flow of information is:

   (a) Media event (b) Media strategy
   (c) Frequenting (d) Gatekeeping

3. The campaign for a New World of Information and Communication Order (NWICO). Was the result of the movement carried out mainly by?

   (a) East European nations (b) African countries
   (c) Nordic countries (d) Non-Aligned Nations

4. The leading financial publication Wall Street Journal has been Acquired by:
   (a) Bill Gates (b) Rupert Murdoch
   (c) Ted Turner (d) Knight-Ridder

5. A specially designed oscilloscope used to graphically display a video signal :
   (a) Vu Meter (b) Viewfinder
   (c) Studio Monitor (d) Waveform Monitor

6. A brief and compact proposal outlining a television programme Production:
   (a) Focus (b) Treatment
   (c) Talent (d) Splice

7. Demographics refer to the examination of the characteristics of The audience such as:
   (a) Attitudes (b) Opinions
   (c) Social and economic (d) Feelings

8. The wire-based international news agencies were the first Significant form of:
   (a) Continental media (b) Regional media
   (c) Global media (d) Sub-regional media

10. Transformational advertising influences:
   (a) Feelings (b) Perceptions
   (c) Learning process (d) Attitudes

11. The modern busy consumer lifestyle results in :
   (a) Positioning (b) Cocooning
   (c) Cognition (d) Hyperbole

12. One of seven major types of appeals in most of the An advertisement is:
   (a) Environment (b) Fear
   (c) Irrationality (d) Submissiveness

13. A theory that suggests that the impact of two or more media is stronger than using Either medium alone is called:

(a) Media message effect (b) Media usage effect
(c) Media survey effect (d) Media multiplier effect

14. Denotative meaning is the common dictionary meaning whereas Connotative meaning Refers to:

(a) Concealed meaning (b) Adverse meaning
(c) Emotional meaning (d) Semantic

15. An effective PR programme requires coordination of all Efforts towards:

(a) Policy making (b) Defined goals
(c) Budgeting (d) Planning

16. The style of news writing in which the identification of the Source is not immediately Revealed:

(a) Summary lead (b) Question lead
(c) Blind lead (d) Satiried lead

17. The process of substantial trimming of the story along with minor facts:

(a) Kill (b) Hole
(c) Slant (d) Boil

18. A set of characteristics that make a brand unique in the Market place:

(a) Follow-up story (b) Wrap-up
(c) Story (d) Highlights

19. The first Indian journalist who took the lead pleading for Free flow of information through a written appeal was:

(a) Tushar Kanti Ghosh (b) Balagangadhar Tilak
(c) Aurobindo Ghosh (d) Rajaram Mohan Roy

20. The degree of clarity in a picture is called as :

(a) Resolution (b) Registration
(c) Shading (d) Sweetening

21. Assertion (a): Often the primary aim of mass media is neither to transmit particular Information nor to write a public in some expression of culture belief Or values, but simply to catch and hold visual or aural attention.

    Reason (r): The message of ritual communication is usually Latent and ambiguous, Depending on associations and symbols which are not chosen by the Participants but made available in culture.

    (a) Both (a) and (r) are true
    (b) Both (a) and (r) are true, but (r) is not the correct explanation
    (c) (a) is true, but (r) is false
    (d) (a) is false, but (r) is true

22. Assertion (a) : Within the space in which reality is played out, the media institutions provide media audiences with information, images, stories and impressions, sometimes according to their own purposes and logic, sometimes guided by other social institutions.

    Reason (r) : Mediation can be a purely neutral process and that it will not have any consistent biases.

    (a) Both (a) and (r) are true
    (b) Both (a) and (r) true, but (r) is not the correct explanation
    (c) (a) is true, but (r) is false
    (d) (a) is false, but (r) is true

23. Assertion (a) : Authoritarian Theory undermines advance censorship and punishment for deviation from rules laid down by political authorities.

    Reason (r) : The theory was likely to be observed in dictatorial regimes, under conditions of military rule of foreign occupation and even during states of extreme emergence in democratic societies.

    (a) Both (a) and (r) are true
    (b) Both (a) and (r) are true, but (r) is not the correct explanation
    (c) (a) is true, but (r) is false
    (d) (a) is false, but (r) is true

24. Assertion (a) : A cut from one shot to another is similar to what our eyes do as they rapidly focus on various parts of our surroundings. The cut can also radically change time and place in the video space.

    Reason (r) : The cut is the most obvious transition because it occurs so slowly and the speed of a cut is not fixed like other transitional devices.

    (a) Both (a) and (r) are true
    (b) Both (a) and (r) are true, but (r) is not the correct explanation
    (c) (a) is true, but (r) is false
    (d) (a) is false, but (r) is true

25. Assertion (a) : Relative to participant receivers, observers represent lesser independence with senders. They offer a distinctive perspective devoid of perpetual bias, cognitive load, relational engagement and conversational demands.

    Reason (r) : Sender participants do not possess greater ability to make discrimination than participant receivers who are occupied with conversational responsibilities.

    (a) Both (a) and (r) are true
    (b) Both (a) and (r) are true, but (r) is not the correct explanation
    (c) (a) is true, but (r) is false
    (d) (a) is false, but (r) is true

26. Assertion (a) : Designing an effective survey study involve aseries of challenging commercial entertainment offerings tend to dominate the media.

    Reason (r) : The market makes money by programming that yieldsnegative externalities that improve bottom line and consequently ignores positivepublic sphere by the media.

    (a) Both (a) and (r) are true
    (b) Both (a) and (r) are true, but (r) is not the correct explanation
    (c) (a) is true, but (r) is false
    (d) (a) is false, but (r) is true

27. Assertion (a) : Newcomb s model implies that any communication system may be characterized by a balance of forces. And any change in any part ofthe system will lead to strain towards balance and symmetry.

    Reason (r) : Because imbalance or lack of symmetry is psychologically uncomfortable and generates internal pressure to restore balance.

    (a) Both (a) and (r) are true
    (b) Both (a) and (r) are true, but (r) is not the correct explanation
    (c) (a) is true, but (r) is false
    (d) (a) is false, but (r) is true

28. Assertion (a) : The experimental filmmakers, whether favouring rhythmical abstraction or surrealistic projections of inner reality, approach the cinema with conceptions which alienate it from nature.

    Reason (r) : They are ignoring the spirit of modern printing and literature and preferring creativity that inhibit camera exploration of reality.

    (a) Both (a) and (r) are true
    (b) Both (a) and (r) are true, but (r) is not the correct explanation
    (c) (a) is true, but (r) is false
    (d) (a) is false, but (r) is true

29. Assertion (a) : The mass media audience is neither organised nor self-acting.

    Reason (r) : Anonymity is the major factor that contributes to this situation.

    (a) Both (a) and (r) are true
    (b) Both (a) and (r) are true, but (r) is not the correct explanation
    (c) (a) is true, but (r) is false
    (d) (a) is false, but (r) is true

30. Assertion (a) : The market finds that violence and sex draw larger audiences, therefore

List-I (Folk Media)		*List-II (State)*	
(a)	Yakshagana	(i)	Jammu and Kashmir
(b)	Jatva	(ii)	Karnataka
(c)	Nautanki	(iii)	West Bengal
(d)	Bhand Pathar	(iv)	Uttar Pradesh

Codes :

	(a)	(b)	(c)	(d)
(a)	(ii)	(iii)	(iv)	(i)
(b)	(i)	(iv)	(ii)	(iii)
(c)	(iii)	(ii)	(i)	(iv)
(d)	(iv)	(iii)	(i)	(ii)

31. Match List-I (Light) with List-II (Function) :

List-I (Light)		*List-II (Function)*	
(a)	Key light	(i)	Lighting from a behind a subject
(b)	Fill light	(ii)	Provides soft light
(c)	Back light	(iii)	Provides hard light
(d)	Scoop	(iv)	Illuminates shadow areas

Codes :

	(a)	(b)	(c)	(d)
(a)	(i)	(ii)	(iv)	(iii)
(b)	(ii)	(i)	(iii)	(iv)
(c)	(iii)	(iv)	(i)	(ii)
(d)	(iv)	(ii)	(i)	(iii)

32. Match List-I (Director) with List-II (Film) :

List-I (Director)		*List-II (Film)*	
(a)	Sergi Eisenstein	(i)	The Bicycle Theives
(b)	D.W. Griffith	(ii)	Nanook of the North

(c) Victoria de Sica	(iii) Battleship Potemkin
(d) Robert Flaherty	(IV) Birth of a Nation

Codes :

	(a)	(b)	(c)	(d)
(a)	(ii)	(iv)	(i)	(iii)
(b)	(i)	(iii)	(ii)	(iv)
(c)	(iv)	(ii)	(iii)	(i)
(d)	(iii)	(iv)	(i)	(ii)

33.

List-I (Author)	*List-II (Book)*
(a) John Milton	(i) Men, Media and Messages
(b) J.S. Mill	(ii) Understanding Media
(c) Marshall McLuhan	(iii) Aeropagitica
(d) Wilbur Sehramm	(iv) On Liberty

Codes :

	(a)	(b)	(c)	(d)
(a)	(i)	(iii)	(ii)	(iv)
(b)	(iii)	(iv)	(ii)	(i)
(c)	(ii)	(iii)	(iv)	(i)
(d)	(iv)	(i)	(ii)	(iii)

34. Match List-I (News agency) with List-II (Country):

List-I (News agency)	*List-II (Country)*
(a) Kyoda	(i) Indonesia
(b) DPA	(ii) Japan
(c) Xinhua	(iii) Germany
(d) Bernama	(iv) China

Codes :

	(a)	(b)	(c)	(d)
(a)	(iv)	(ii)	(i)	(iii)
(b)	(i)	(iii)	(ii)	(iv)
(c)	(iii)	(i)	(ii)	(iv)
(d)	(ii)	(iii)	(iv)	(i)

35. Match List-I (Radio) with List-II (Owner):

List-I (Radio)	*List-II (Owner)*
(a) Big FM	(i) Times of India group
(b) Radio Citi	(ii) NDTV group
(c) Radio Mirchi	(iii) Star group
(d) Red FM	(iv) Reliance group

Codes :

	(a)	(b)	(c)	(d)
(a)	(i)	(iii)	(ii)	(iv)
(b)	(iv)	(iii)	(i)	(ii)
(c)	(iii)	(ii)	(iv)	(i)
(d)	(ii)	(iii)	(i)	(iv)

36.

List-I (Director)	*List-II (Film)*
(a) Guru Dutt	(i) Do Bigha Zameen
(b) Bimal Roy	(ii) Abhimaan
(c) Mehboob Khan	(iii) Pyaasa
(d) Hrishikesh Mukherji	(IV) Mughul-e-Azam

Codes :

	(a)	(b)	(c)	(d)
(a)	(i)	(iv)	(iii)	(ii)
(b)	(iv)	(ii)	(iii)	(i)
(c)	(iii)	(i)	(iv)	(ii)
(d)	(ii)	(i)	(iii)	(iv)

37. Match List-I (Country) with List-II (Newspaper):

List-I (Country)	*List-II (Newspaper)*
(a) France	(i) AL HA-RAM
(b) Pakistan	(ii) The Times
(c) Egypt	(iii) Le Monde
(d) England	(iv) The Dawn

Codes :

	(a)	(b)	(c)	(d)
(a)	(iii)	(iv)	(i)	(ii)
(b)	(i)	(iv)	(ii)	(iii)

(c)	(ii)	(iii)	(iv)	(i)
(d)	(iv)	(i)	(ii)	(iii)

38. Match List-I (Person) with List-II (Field):

List-I (Person)		*List-II (Field)*	
(a)	Leonal Fielden	(i)	Radio
(b)	Noam Chomsky	(ii)	Advertising
(c)	David Ogilvy	(iii)	Public Relations
(d)	Ivy Lee	(iv)	Media Criticism

Codes :

	(a)	(b)	(c)	(d)
(a)	(ii)	(iv)	(i)	(iii)
(b)	(iii)	(i)	(iv)	(ii)
(c)	(iv)	(iii)	(ii)	(i)
(d)	(i)	(iv)	(ii)	(iii)

39. Identify the correct chronological order of films:
    (a) Guddi - Bhoomika - Rajniganda - Kaagaj ke Phoole
    (b) Kaagaj ke Phoole - Guddi - Rajniganda - Bhoomika
    (c) Rajniganda - Guddi - Kaagaj ke Phoole - Bhoomika
    (d) Bhoomika - Guddi - Kaagaj ke Phoole - Rajniganda

40. Identify the correct sequence of evolution of printing press :
    (a) Rotary - Cylinder - Treddle - Offset
    (b) Cylinder - Rotary - Offset - Treddle
    (c) Treddle - Cylinder - Rotary - Offset
    (d) Rotary - Treddle -Offset - Cylinder

41. Identify the correct sequence of research in content analysis:
    (a) Data analysis - subject categories - unit of analysis - coding
    (b) Unit of analysis - subject categories - coding - data analysis
    (c) Unit of analysis - subject categories - data analysis - coding
    (d) Unit of analysis - data analysis - coding - subject categories

42. Find the correct sequence of TV channels when they were established in India:

    (a) Zee TV - Star World - Sony TV - Discovery Channel
    (b) Zee TV - Sony TV - Discovery Channel - Star World
    (c) Zee TV - Star World - Sony TV - Discovery Channel
    (d) Sony TV - Zee TV - Star World - Discovery Channel

43. Identify the correct sequence of evolution of video camera:

    (a) Hi 8 - VHS - Digital - Betacam
    (b) Hi 8 - Digital - VHS - Betacam
    (c) VHS - Hi 8 - Betacam - Digital
    (d) VHS - Digital - Hi 8 - Betacam

44. Identify the correct order of research elements from the following:

    (a) Variable - Construct - Concept - Hypothesis
    (b) Hypothesis - Variable - Concept - Construct
    (c) Concept - Construct - Variable - Hypothesis
    (d) Construct - Hypothesis - Concept - Variable

45. Which date is very important for parliamentary privileges in India?

    (a) 15 August 1947 (b) 30 January 1948
    (c) 26 January 1950 (d) 24 May 1964

❑❑❑

Glossary

A

Acculturation. A process of cultural transformation initiated by contacts between different cultures. At a global level, acculturation takes place as societies experience the transforming impact of international cultural contact. The global trend towards modern economic organization and developed market economies has been accompanied by a process of cultural transformation. Individuals experience acculturation when their social roles and socialization are shaped by norms and values that are largely foreign to their native culture.

Active listening. Process of analyzing and evaluating what another person is saying in an effort to understand the speaker's feelings or the true meaning of the message.

Active public. People who are aware of a problem and will organize to do something about it.

Ad hominem. A fallacy that attacks a person rather than the argument itself. This is also referred to as "name calling."

Agenda. List of all topics to be discussed during a meeting.

Agenda-setting. Function of mass media to the relative importance of our attitudes on issues. The perceived importance of issues is related to the attention given to those issues by the media.

Agenda-setting effect. The influence of the mass media created by emphasizing certain topics, thus causing people to perceive these same issues as important.

Artifact. Ornament or possession that communicates information about a person.

Asynchronous communication. Also known as non simultaneous communication. Communication during which participants engage in the process at different times, such as bulletin boards or e-mail.

Attitude(s). Learned predisposition to respond favorably or unfavorably toward an object. Stable clusters of feelings, beliefs, and behavioral intentions toward specific objects, people, or institutions. A construct said to be composed of affective (feeling), cognitive (thoughts), and connotative (behavioral) components; internal feelings about some object; composed of opinions, beliefs, and values. Evaluative disposition, feeling, or position about oneself, others, events, ideas, or objects.

Audience. A group of individuals attending to a common media. They receive communication from the same source, but are not active participants and do not communicate with each other. Collection of individuals who have come together to watch or listen to someone or sething, such as to listen to a speech.

Authoritarian leadership style. The leader controls group goals and procedures.

Autocrat personality. A person who possesses a need to dominate others and distrusts the motives and creative potential of others.

Avoiding style. An approach to conflict which tries to manipulate circumstances so that conflict does not surface.

Aware public. People who know about a problem but don't act on it.

B

Belief. An expectation about the way some event or sequence of events will occur. Conviction or confidence in the truth of some-thing that is not based on absolute proof.

Benchmarking. The process of seeking to improve quality by comparing one´s own products or services with the best products or services of others.

Brainstorming. A technique designed to foster group productivity by encouraging interacting group members to express their ideas in a noncritical fashion.

Bureaucracy. Organization characterized by hierarchical chains of command and power, each with its own separate function. The organization is governed by rules accepted by the members in order for efficient mass administration. A formal organization with defined objectives, a hierarchy of specialized roles and systematic processes of direction and administration. Bureaucracy is found in earlier times in history, for example in administration of agricultural irrigation systems, the Roman army, the Catholic church, but it becomes most prominent in the large-scale administration of agencies of the modern state and modern business corporations.

C

Campaign. In advertising, a large number of ads that stress the same theme and appear over a specified length of time.

Centralization. The degree to which information must flow through a specific central member of a communication network.

Change agent. Individual who exerts influence on opinion leaders to adopt an innovation.

Charisma. An attitude of enthusiams and optimism that is contagious; an aura of leadership. The appeal or attractiveness that the audience perceives in the speaker, contributing to the speaker's credibility. [SB] Channel. Medium or route through which a message is sent for purposes of communication. The pathway by which a message travels from sender to receiver. Means by which a message is conveyed from source to receiver (radio, television, telephone, face-to-face, for example). Route (such as sound waves or light waves) by which messages flow between the source and the receiver.

Cheesecake. Photograph of a scantily clad young woman used as a publicity device. Similar photographs of men are called "beefcake".

Chronemics. The study of how people perceive, structure, and use time as communication.

Code. Set of rules or symbols used to translate a message from one form to another.

Collaborative style. A problem-solving approach to conflict situations where consulting with affected parties is considered important.

Commitment. Desire of group members to work together to complete a task to the satisfaction of the entire group.

Communication. The mutual process through which persons interpret messages in order to coordinate individual and social meanings. Human manipulation of symbols to stimulate meaning in other humans. The proces by which a person, group, organization (the sender) transmits some type of information (the message) to another person, group, organization (the receiver). The simultaneous sharing and creating of meaning through human symbolic action.

Communication apprehension. Fear or anxiety associated with real or anticipated communication with others. Anxiety syndrome associated with either real or anticipated communication with another person or persons.

Communication competence. Ability to take part in effective communication, which is characterized by skills and understandings that enable communication partners to exchange messages successfully.

Communication flow. The direction (upward, downward, horizontal) messages travel through the networks in an organization.

Communication networks. The patterns of communication flow between individuals in organization. Pre-established patterns dictating who may communicate with whom.

Communication policies. Final statements of organizational positions related to communication activities and behaviors and information sharing.

Communitarian. A philosophy or belief system which places priority on the community or on social values. Often contrasted to individualism or libertarianism. It claims that meaning in individual life and individual liberty are only possible within a strong and vital community so government policies and individual choices should be responsive to social values.

Community. A society where peoples relations with each other are direct and personal and where a complex web of ties link people in mutual bonds of emotion and obligation. In the social sciences, especially sociology, the idea of community has provided a model to contrast to the emergence of more modern less personal societies where cultural, economic and technological transformation has uprooted tradition and where complexity has created a less personal and more rationalized and goal-directed social life.

Competitive style. A win/lose approach to conflict situations which typically involves a good deal of argumentation.

Compromising style. An approach to conflict which emphasizes all parties getting some of what they want.

Computer-Mediated Communication . Any form of interpersonal, small group, organizational, or puclic communication that occurs with the use of computers.

Conduit metaphor. The persistent bias within the English language toward assuming that communication is a transmission of meanings "contained" in words from "senders" to "receivers."

Constructivism. Theory explaining communication as a fundamentally interpretive process through which persons organize and create personal reality; examines the complexity with which persons develop and organize different categories of perception called constructs.

Contentious style. Tendency to challenge others when disagreements occur.

Controlled media. Those media that the public relations practitioner has actual control over, such as a company newsletter.

Cool media. McLuhan´s term for media that are incomplete and, thus, require human involvement and participation.

Copyright. The protection of a creative work from unauthorized use. Legal protection from unauthorized use of intellectual property fixed in any tangible medium of expression.

Corporate communications. Term covering all types of communication by a company to both external and internal audiences.

Corporate image. The impression that people have of an organization.

Corporate stories. Narratives that serve to convey the values, style, and history of an organization. Told to newcomers, stories perform socialization; told among veteran members of an organization, stories serve to bind members together and vitalize the organization´s ideology.

Counterculture. A set of cultural ideas that, to some extent, differ from and conflict with, those generally upheld in the society. A counterculture develops when members of groups identify common values that distinguish them from others. These groups may be based on common appearance, ethnic group, sexuality, status or social behaviour. The term is close in meaning to subculture, but the concept of counterculture stresses the idea of an open and active opposition to dominant cultural values.

Credibility. Speaker's believability, based on the audience's evaluation of the speaker's competence, experience, character, and charisma.

Crisis communications. Methods and policies a corporation uses in distributing information when its operations become involved in an emergency situation affecting the public.

Critical analysis. Form of research that goes beyond description and explanation to argue for changes in communicative practices that are judged to be oppressive, wrong, or otherwise undesirable.

Critical theory/ies. Explains and critiques social communication by examining the implications of power relationships; Marxist and feminist scholars usually approach theorizing with these assumptions. Often identified with cultural studies approaches. Group of theories that seek to produce change in oppressive or otherwise undesirable practices and structures in society.

Critical listening. Listening that judges the accuracy of the information presented, determines the reason-ableness of its conclusions, and evaluates its presenter.

Cross-cultural communication. Communication of different combinations of people. A cross-cultural communication study might compare and contrast Japanese and American negotiation tactics, for example.

Cultivation. Cumulative process by which television fosters beliefs about social reality including the belief that the world is more dangerous and violent than it actually is.

Cultivation analysis. Explains media, especially television, not in cause?effect terms but by describing longer?term tendencies of audiences to adjust their expectations about reality in the direction of prevalent media content.

Cultivation theory. Point of view that claims television cultivates, or promotes, a view of social reality that may be inaccurate, but that viewers nonetheless assume reflects real life.

Cultural imperialism. The practice of systematically spreading the influence of one culture over others by means of physical and economic domination. Usually involves an assumption of cultural superiority (ethnocentrism).

Cultural mainstream. General view of life in a society. Cultivation theorists argue that television constructs and presents images that define the cultural mainstream.

Cultural studies. Broadly based approach to studying communication through its cultural implications; often studies in this tradition adopt the tone of social commentary or criticism; closely identified with critical theory. Group of related theories that seek to unmask the techniques by which privileged groups maintain their privilege and power in society.

Culture. Traditions and patterns of thought which are passed down through generations of people. The generally shared knowledge, beliefs and values of members of society. Culturis conveyed from generation to generation through the process of socialization. The set of values, customs, and beliefs that puople have in common with other members of a social unit (e.g., a nation). The deposit of knowledge, experience, beliefs, values, actions, attitudes, meanings, hierarchies, religion, notions of time, roles, spatial relations, concepts of the universe, and artifacts acquired by a group of people in the course of generations through individual and group striving. Structures and practices that uphold a given social organization by producing and reproducing particular values, expectations, meanings, and patterns of thought, feeling, and action. Both the ideology of a society and the actual, concrete practices that occur in that society.

Culture shock. The tendency for people to become confused and disoriented as they find it difficult to become adjusted to a new culture.

D

Decentralization. The extent to which authority and decision makng are spread throughout all levels of an organization rather than being reserved for top management (centralization).

Decentralized networks. Communication networks in which all members play an equal role in the transmittal of information.

Decoding. Process of translating a message into the thoughts or feelings that were communicated.

Defensive communication. Behavior which occurs whenwhen a person perceives or anticipates threat in interaction.

Democratic leadership style. The leader seeks group member participation in determining group goals and procedures.

Democrat personality. A person who can give or receive orders, follow or lead, and is capable of relinquishing control.

Dominant style. Tendency to lead and take control in social situations.

Downsizing. The process of adjyusting downward the number of employees required to perform jobs in newly designed organizations.

Downward communication. Communication from higher members of the organization (i.e., managers, vice-presidents) to members lower in the organizational hierarchy.

Dramatistic pentad. Means of analyzin rhetoric in context through looking at the five factors (pentad) of act (what was done), agent (by whom it was done), scene (where it was done), agency (by what means it was done), and purpose (the goal that guided the action).

Dyad. A two-person communication system.

E

Editor. Director of a newspaper´s news and editorial department; may be subordinate to the publisher or on equal footing, depending upon the newspaper´s organization.

Egalitarian. A shortening of the word equalitarian, suggesting a commitment to, or a state of, equality. Egalitarian societies or groups are contrasted to hierarchical or class-based societies or groups.

Electronic epoch. Fourth era in media history of civilization. The electronic epoch was ushered in by the invention of the telegraph, which made it possible for people to communicate personally across distance.

Empathic listening. Listening to understand what another person is thinking and feeling. Empowerment. The passing of responsibility and authority from managers to employees.

Encoding. Process by which the source expresses thoughts or feelings in words, sounds, and physical expressions, which together make up the actual message that is sent.

Ethnocentrism / ethnocentric. Judging other groups according to the categories and values of one´s own culture rather than being open to cultural differences. The assumption that the culture of one's own group is moral, right and rational and that other cultures are inferior. When confronted with a different culture, individuals judge it with reference to their own standards and make no attempt to understand and evaluate it from the perspective of its members. Sometimes ethnocentrism will be combined with racism, the belief that individuals can be classified into distinct racial groups and that that there is a biologically-based. A person whose pride in his or her her-itage or background leads to the conviction that he or she knows more and is better than those who differ.

Ethos. Aristotelian concept associated with persuasion; the personal character of the speaker.

Evaluative listening. Listening to judge or analyze information.

Expatriates. People who are citizens of one country but who are living in another country.

F

Fallacy. Arguments that are flawed because they do not follow the rules of logic.

Feedback. Any message that aids a communicator in evaluating the success of previous message. The responses of the receiver that shape and alter subsequent messages from the source.

Feminine communication orientation. Fostered by socialization in feminine communication cultures, this orientation assumes that the purposes of communication are to create and sustain relationships, support others, and maintain harmony between people. It is enacted through communication that is cooperative, supportive, emotionally expressive, relationship-focused, and attentive to interactional processes.

Feminine culture. According to Hofstede, the cultural orientation in which people emphasize concern for others and the relationship among people. [Feminist theory. Explains communication variables such as language, nonverbal immediacy, media effects, and ethics from the perspective of gender relationships; often focuses on overt and covert power implications of cultural patterns.

Flow. Stream of communication messages.

Fordism. Refers to the system of mass production (e.g., the assembly line) pioneered by Henry Ford to meet the needs of a mass market.

Formal communication systems. Communication links and networks determined and sanctioned by the organization.

Formal groups. Groups that are created by the organization, intentionally designed to direct its members toward some organizational goal. [Format. Consistent programming designed to appeal to a certain segment of the audience.

Function of communication. According to Cicero, to entertain, inform, and persuade: to stimulate was added later.

Functional approach to leadership. The study of leadership which focuses on the leadership behaviors needed by the group to accomplish its goals.

G

Gangplank. Horizontal chain of communication between employees on the same hierarchical level but in different departments.

Gatekeeper. Any person (or group) who controls what media material eventually reaches the public. Editor, reporter, news director, or other person who decides what material is printed, broadcast, or otherwise offered to the public. Individual who controls the flow of information to a group of people. An individual who is positioned within a communication network so as to control the messages flowing through communication channels. A filter between source/ receivers in the mechanistic model of communication.

Gemeinschaft. A German word, translated as 'community', used by sociologist Ferdinand Tonnies to define an 'ideal type', or model, society where social bonds are personal and direct and there are strong shared values and beliefs. Characteristic of small scale, localized societies, it is in contrast to Gesellschaft which refers to complex, impersonal societies.

Gender. A social construct related to masculine and feminine behaviors that are learned. A social, symbolic construction that expresses the meanings a society confers on biological sex. Feminine and masculine are gender terms that refer to socially expected and prescribed qualities in women and men, respectively. Sex and gender are not absolutely correlated, so women and men may be masculine or feminine or degrees of both. Socially created system of values, identities, and behaviors that are prescribed for women and men. Unlike sex, which is biologically determined, gender is socially constructed.

Gender roles. Social roles ascribed to individuals on the basis of their sex. The term gender differs from sex because it refers specifically to the cultural definition of the roles and behaviour appropriate to members of each sex rather than to those aspects of human behaviour that are determined by biology. Thus giving birth is a female sex role, while the role of infant nurturer and care giver (which could be performed by a male) is a gender role usually ascribed to females.

General Systems Theory. The description of living systems in terms of the interdependence of their components and relationships among components. A view of systems that examines the system as more than the simple interconnectedness between objects; presupposes an active and reactive system fed on information.

Gesellschaft. A German word, translated as 'society-association', used by Ferdinand Tonnies to refer to an 'ideal type', or model, of a society where social bonds are primarily impersonal, instrumental and narrow. Characteristic of large scale, complex societies, with a strict division between private and public spheres of life, it contrasts to the community-oriented life of the Gemeinschaft.

Glass ceiling. A barrier preventing females from reaching top positions in many organizations. In the analysis of women in the work place, this concept is useful for describing the invisible barriers that block the promotion of women. It refers to barriers that are not explicit, but are inherent in the social organization and social relationships of the workplace. For example, women may find their corporate careers obstructed because they are excluded from the recreational and social associations created by male fellow workers and lack the social contacts that are important in gaining status and recognition.

Globalization. The process of interconnecting the world's people with respect to the cultural, economic, political, technological, and environmental aspects of their lives. A comprehensive world-wide process of the internationalisation of communication, trade and economic organization. In the economic sphere it can be seen in international trade agreements, vast increases in the volume of international trade and growing economic interdependency. It is also marked by the expansion of the size and power of multinational corporations and the development of the American entertainment industry's domination of international cultural communication. Generally the process is seen as driven by the growth of international capitalism and involving the transformation of the culture and social structures of non-capitalist and pre-industrial societies.

Grapevine. An organization's informal channels of communication, based mainly on friendship or acquaintance, Grass roots lobbying. Organizing local constituencies to influence government decision makers.

Great person theory. The view that leaders possess special traits that set them apart from others and that these traits are responsible for their assuming positions of power and authority.

Group. A collection of two or more interacting individuals who maintain stable patterns of relationships, share common goals, and perceive themselves as being a group. An aggregate of individuals having some characteristic in common. They may be distinguished from others by appearance, language, socio-economic status or cultural values and practices. A group is often characterized by a sense of common identity, shared interests and goals among its members, but a group may exist simply because its members share some objective characteristic and are defined as a group by others. Collection of individuals who form a system in which members influence one another, derive mutual satisfaction from one another, have a common purpose, take on roles, are interdependent, and interact with one another.

Group Decision Support System. A computer-based system that supports message exchange, collaboration on ideas, projects, and products, and/or group decision-making.

Groupthink. A group communication process where high group cohesiveness impairs decisions by stimulating premature closure on important issues. The tendency for members of highly cohesive groups to conform to group pressures regarding a certain decision so strongly that they fail to think critically, rejecting the potentially correcting influences of outsiders. A dysfunction in which group members value the harmony of the group more than new ideas, fail to critically examine ideas, hesitate to change de-cisions, or lack willingness to allow new members to participate.

H

Halo effect. The tendency for our overall impressions of others to affect objective evaluations of their specific traits; perceiving high correlations between characteristics that may be unrelated.

Haptics. Tactile, or touch, communication; one of the most basic forms of communication.

Hawthorne effect. An increase in worker productivity observed at the Chicago Hawthorne plant of General Electric in the 1920's and 1930's attributed to improvements in worker-management communication and increased involvement of workers with each other. The term is now used more generally to refer to improvement of worker productivity that does not result from any objective change in working conditions or work organization, but seems to arise from workers having more positive psychological feelings about the workplace.

Hermeneutics. The branch of philosophy that investigates the interpretation of texts; in one form popularized by Gadamer, hermeneutics emphasizes the historical and inherently linguistic nature of experience in denying a transmissional model of communication.

Hierarchy. Ordering in which parts are related to each other in subordinate or superordinate fashions. For example, they may be more or less important, large or complex. A system is composed of a hierarchy consisting of subsystems and suprasystems. A proposition underlying systems theory which maintains that systems are organized in a successively more inclusive and complex pattern and that to understand systems of behavior, several appropriate levels should be examined. A structuring of social statuses and roles within an organization or society ranked according to differentiations of power, authority, wealth, income, etc. Related terms are ranking or stratification.

High-context culture. Culture in which most of the information in a message is encoded in the physical context or in the person's mental catalog of rules, roles and values. [The meaning of the communication act is inferred from the situation or location.

Homeostasis. A characteristic of systems whereby feedback seeks to maintain the system at the current level. A steady state, equilibrium, balance. General systems theory claims that living systems (relationships, for example) strive for, but never fully achieve, homeostasis. Dialectical theory, on the other hand, claims that continuous change is the very nature of relationships.

Horizontal chain of communication. Communication between organization. members on the same hierarchical level (between two managers or between two subordinates, for example).

Hot media. McLuhan's term for relatively complete media that do not require significant human participation.

Human capital. The talents and capabilities that individuals contribute to the process of production. Companies, governments and individuals can invest in this 'capital' just as they can invest in technology and buildings or in finances.

Humanism. Form of science that focuses on human choices, motives, and meanings and that assumes the reasons or causes of human behavior lie within humans, not outside of them.

Hypodermic needle theory. The belief that people receive information directly without any intervening variable, as in a vacuum.

I

Immediacy. Theory of immediacy demonstrates how persons signal the emotional responses of attraction, dominance, and arousal through nonverbal (and verbal) messages related to physical closeness; for example, eye contact and movement toward another, as indicators of immediacy, correlate with desires for increased involvement.

Impression management. Efforts by individuals to improve how they appear to others. Creating a positive image of oneself in order to influence the perceptions of others.

Informal communication systems. Communlcation links and networks (not determined by the organizatinal chart) which arise through natural human interaction. For example, two workers who might have no formal communication links may be

connected in the informal communication system because they both play on the company golf team or eat lunch together.

Informal groups. Groups that develop naturally among people, without any direction from the organization within which they operate.

Interaction. Exchange of communication in which communicators take turns sending and receiving messages.

Intercultural communication. Communication between individuals or groups from different cultures or from different subcultures (for example, ethnic groups) of the same sociocultural system. Branch of communication field that studies communication between people from different cultures, including distinct cultures within a single country.

Interest group. A group of individuals and organizations linked together for the purpose of active promotion of particular values and objectives. Interest groups are usually associated with the political process through which they seek support and resources for their objectives.

Internet. Master computer network connecting networks world-wide, enabling computer users to exchange e-mail, hold electronic conversations, obtain information and entertainment, and operate sites on the World Wide Web.

Interorganizational communication. Structures communication among organizations linking them with their environments.

Interpersonal communication. Exchange between two or more persons in close proximity using conversation and gestures. The exchange of messages between individuals through which needs, perceptions, and values are shared and by which mutual meanings and expectations are developed. Communication between individuals. Interpersonal communication exists on a continuum ranging from impersonal (between social roles) to highly personal.

Interpretation. An active process whereby individuals perceive and assign meaning to phenomena, relying on their working models to do so.

Interview. Carefully planned and executed question-and-answer session designed to exchange desired information between two parties. Intrapersonal communication. Communication with oneself, including self?talk, planning, and reflections.

Issues management. Program of identifying and addressing issues of public concern in which a company is or should be involved. The process of identifying issues that potentially impact organizations and managing organizational activities related to those issues.

K

Key contacts. People who can either influence the publics an organization is trying to reach or who have direct power to help the organization.

Kinesics. Sometimes referred to as "body language"; any movement of the face or body that communicates a message.

K.I.S.S principle. A basic principle of communication advising that messages should be as short and simple as possible (an abbreviation for keep it short and simple).

L

Laissez-faire leader. Leader who gives complete decision-making freedom to the group or to individual members.

Language. Structured system of signs, sounds, gestures, and marks used and understood to express ideas and feelings among people within a community, nation, geographic area, or cultural tradition. [Laissez-faire leadership style. The leader provides the means for members to accomplish group goals and has minimal involvement in group functioning.

Latent public. People who are not aware of an existing problem.

Leadership. The process whereby one individual influences other group members toward the attainment of defined group or organizational goals. An influence process that includes any behavior that helps clarify or guide the group to achieve its goals.

Liaison(s). Person who links two groups but is not a member of either group. Individuals who serve as linking pins connecting two or more groups within organizational communication networks. Sometimes referred to as internal boundary spanners.

Libertarianism. A philosophy or belief system which gives priority to the liberty of the individual. May be associated with classical liberalism regarding economic matters or the protection of those negative liberties which declare the right of the individual to be free from interference by the state, or the community, unless the actions of the individual constitute harm to others. For example, the individual has the right to freedom of speech, freedom of association, freedom of religious expression, freedom of contract. Libertarianism is related to individualism and contrasted with communitarianism.

Line organization. A method of structuring organizations as a sequence of ascending levels of responsibility for the production of goods or services.

Listening. The active process of receiving aural stimuli by hearing, selecting, attending, understanding, evaluating, and remembering.

Literate epoch. Second era in media history of civilization. Invention of the phonetic alphabet inaugurated the literate epoch in which common symbols allow people to communicate with writing.

Lobbying. The practice of trying to influence governmental decisions. Usually done by agents who serve interest groups.

Lobbyist. Person who presents an organization´s point of view to members of government bodies.

Logos. Aristotelian concept associated with persuasion; proof or apparent proof provided by the words used in the speech.

Low-context culture. Culture in which most information in a message is contained in the explicit or verbal message. The meaning of the communication act is inferred from the messages being sent and not the location where the communication occurs.

M

Machiavellianism. Personality trait which involves manipulating other people as a basic strategy of social influence. A personality trait involving willingness to manipulate others for one's own purposes.

Magic bullet theory. Influential early perspective on media effects; held that media caused direct and measurable effects on individuals in the mass audience. Theory which suggests that the mass media influence a large group directly and uniformly. Also referred to as the hypodermic needle theory.

Mainstreaming. Argues that heavy television viewing diminishes differences in perceptions of reality caused by demographic and social factors. The effect of television in stabilizing and homogenizing views within a society.

Management-by-objectives (MBO). A process that specifies that supervisors and employees will jointly set goals for employees. Usually followed by a joint evaluation of the employee's progress after a set period of time. The technique by which managers and their subordinates work together to set andhen meet organizational goals.

Mass communication. The process by which a complex organization, with the aid of one or more machines, produces and transmits public messages that are directed at large, heterogeneous, and scattered audiences. Communication to large audiences which is mediated by electronic or print media.

Marketing communications. Product publicity, promotion, and advertising.

Masculine communication orientation. Cultivated by participation in masculine communication cultures, this orientation views the goals of communication as asserting self, competing for attention and influence, and achieving objectives. It is enacted through a communication style that is competitive, individualistic, emotionally reserved, and instrumental.

Masculine culture. According to Hofstede, cultures in which people are highly materialistic and value assertiveness and the acquisition of money.

Mass culture. A set of cultural values and ideas that arise from common exposure of a population to the same cultural activities, communications media, music and art, etc. Mass culture becomes possible only with modern communications and electronic media. A mass culture is transmitted to individuals, rather than arising from people's daily interactions, and therefore lacks the distinctive content of cultures rooted in community and region. Mass culture tends to reproduce the liberal value of individualism and to foster a view of the citizen as consumer.

Mass media. The channels of mass communication. Sociologically speaking in modern times the 'community' has been replaced by a 'mass', a set of autonomous and disconnected individuals, with little sense of community. The mass media then is that media (radio, television, newspapers, etc) which are targeted at the mass rather than at specific groups or communities.

Matrix organization. The type of departmentalization in which a product or project form is superimposed on a functional form.

Mcworld. A concept developed by Benjamin Barber to describe the new globalized world where nation states have little power and citizenship has become meaningless as a cornerstone of democracy. This new world is ruled by corporations (multinational corporations or in Barber's terms antinational corporations) which see everyone simply as consumers. In this new world citizens can no longer effectively use democracy to enhance or protect social values because this would interfere with the marketplace. The assumption is that the actions of countless consumers will best satisfy the social needs of communities.

Meaning. A human construction arising out of interpreting and negotiating interpretations with others. *Mechanistic organization.* An organizational structure in which people perform specialized jobs, many rigid rules are imposed, and authority is vested in a few top-ranking officials.

Media richness. The degree to which a medium facilitates feedback or provides multiple cues to reduce message ambiguity. Rich media are considered most efficient for highly ambiguous communication.

Mediated interpersonal communication. Any situation where mediated technology (for example, telephone, computer) is used to advance face-to-face interaction.

Medium is the message. A central idea of communications theorist Marshall McLuhan who demonstrated that each media (print, speech, television) is connected with a different pattern or arrangement among the senses and thus results in a different awareness or perception. Although the literal message of a radio report of a disaster and the television coverage of the same event may be identical, the event will be perceived differently and take on different meaning because the two media arrange the senses differently. In this sense the medium (the singular for the word media) is the message; this message is often more important than the literal message.

Message. A stimulus to which meanings are attributed in communication. Set of verbal and/or nonverbal symbols sent to a receiver.

Metacommunication. The process of communicating about communication.

Metaphor. A figure of speech in which a word or phrase relates one object or idea to another object or idea that are not commonly linked together.

Model. A verbal or pictorial description or representation of a process. A way of looking at something. A representation of something else. Models may represent their referents physically, verbally, and/or visually.

Moderating public. Those people who could make it easier for an organization to get its message through to the public it really wants to reach.

Monitor. Process of observing ourselves and our actions. Monitoring is possible because humans are self-reflective.

MUM effect. The reluctance to transmit bad news, shown either by not transmitting the message at all, or by delegating the task to someone else.

N

Narrative paradigm/theory. Point of view that asserts humans are natural storytellers and that most, if not all, communication is storytelling.

News conferences. Structured opportunities to release news simultaneously to all media.

News release. A story prepared for the media to share information and generate publicity.

Newsletters. Regiularly published internal documents describing information of interest to employees regarding an array of business and nonbusiness issues affecting them.

Noble self. Tendency to be inflexible in expressing a position; to base behavior on a rigid conception of self.

Noise. Any internal or external interference with the sending and receiving of messages.

Nonverbal communication. The transmission of messages without the use of words (e.g., by gestures, the use of space). One of two major communication code systems; sometimes defined as all that language is not; communicated via channels other than words. Any information that is expressed without words.

Norm. A culturally established rule prescribing appropriate social behaviour. Norms are relatively specific and precise and elaborate the detailed behavioural requirements that flow from more general and overarching social values . For example, it is a value in Western society that one should respect the dead, it is a norm that one should dress in dark colours for a funeral.

Not-for-profit organization. A group or company whose primary purpose is not to make a profit, regardless of whether it actually does so in a given year.

O

Opinion. The verbal or nonverbal expression of an attitude.

Opinion-leader(s). Person who influences the opinions, attitudes, beliefs and behaviors of others through informal communication. People who are instrumental in influencing other people's attitudes or actions.

Organization. Hierarchically organized group of people so large that personal relationships with every member of the group are impossible. Organizations tend to outlive individual members and to be regulated by formal structures and rules. [IRW] A structured social system consisting of groups and individuals working together to meet some agreed-upon objectives.

Organizational change. Planned or unplanned transformations in an organization's structure, technology, and/or people.

Organizational chart. A diagram representing the connections between the various departments within an organization; a graphic representation of organizational structure, indicating who is to communicate with whom.

Organizational climate. The collective subjective perceptions held by an organization's employees concerning organizational policies, structure, leadership, standards, values, and rules.

Organizational commitment. The extent to which an individual identifies and is involved with his or her organization and/or is unwilling to leave it.

Organizational communication. Communication between and among the indlviduals and groups which make up an organization. The exchange and interaction of informal and formal messages within networks of interdependent relationships.

Organizational culture. A cognitive framework consisting of attitudes, values, behavioral norms, and expectations shared by organization members. Understandings about identity and codes of thought and action that are shared by members of an organization.

Organizational structure. The formal configuration between individuals and groups with respect to the allocation of tasks, responsibilities, and authorities within organizations. The formally prescribed pattern of interrelationships existing between the various units of an organization.

Outsourcing. The practice of eliminating nonessential aspects of business operations by hiring other companies to perform these tasks.

P

Paralanguage. Vocal (but nonverbal) dimension of speech; the manner in which something is said rather than what is said. The way we vocalize, or say, the words we speak.

Parasocial relationship. The tendency of some audience members to identify with media figures (such as celebrities or fictional characters) as though an interpersonal relationship has been established.A situation whereby audience members develop a sense of kinship or friendship with media personalities. A type of "relationship" which exists between television viewers and media performers (for example, talk show hosts, entertainers, sports stars).

Pathos. Aristotelian concept associated with persuasion; the emotive aspects of the speech and audience.

Perception. The process by which an organism assimilates, interprets, and uses sensory data.

Personal relationship. Voluntary commitments that are marked by continuing and significant interdependence between particular individuals and that are constantly in process.

Personalized communication. Using nicknames, coded terms, and special vocabulary to enhance partners' feelings of being connected and to exclude others by demarcating the boundaries of an intimate relationship.

Persuasion. Attitude change toward a source's proposal resulting from a message designed to alter a receiver's beliefs about the proposal. Communication process, involving both verbal and nonverbal messages, that attempts to reinforce or change listeners' attitudes, beliefs, values, or behavior.

Plagiarism. Use of another person's information, language, or ideas without citing the originator and making it appear that the user is the originator.

Planned publicity. Publicity that is the planned result of a conscious effort to attract attention to an issue, event, or organization.

Policy. A type of standing plan that serves as a guide for decision making and usually is set by top management.

Popular culture. Intellectual opinions of popular culture, the culture of the masses, have been deeply shaped by critical theory. Since the Frankfurt School, which identified with the 'high culture' of the intellectual classes, popular culture has been seen as trivial, demeaning and commercialized, serving the interests of the capitalist system. Post-modernist theorists, however, no longer accept the belief that there is some objectively superior high culture setting a standard from which to make evaluations of others. They have been more interested in popular culture as representing the voices of the previously silent, and by adopting the methods of film analysis or literary criticism they examine the way popular culture is produced and the underlying assumptions upon which its meaning rests.

Positioning. The practice of creating corporate identity programs that establish a position in the market for a company and its products. Also, the effort to get ahead by doing something first.

Positivism. One way to think about the relationship between science and society and found in the early writings of August Comte. All of the assumptions that Comte makes are now rejected by postmodernists. Comte begins by imposing meaning on history, arguing that societies evolve through three stages: the theological stage, the metaphysical stage and the positive (or scientific) stage. Each of these stages is reproduced in the evolution of the human mind. The human mind, and the most privileged among these was the sociologist, would use the scientific method to arrive at an understanding of the universal laws of social development. Comte argues against democratic discourse in the belief that parties involved in the political process are always committed to a particular viewpoint. Only science can rise above the local and particular and understand impartially. The application of this knowledge to society would enable the liberation of individuals. Positivism, therefore, placesscience in a privileged position; assumes the possibility of a scientific understanding of human and social behaviour; assumes the separation of knowledge and power; and assumes the possibility of objectivity and impartiality. Positivism shaped sociology for the next 100 years. In much contemporary social science debate, however, positivism has become a term of abuse.

Postmodern. Refers to era of social life that emerged after modernism. Postmodern society is described as fragmented, uncertain, and continuously in flux; individuals are described not as a core self, but as a range of selves that are brought forth by and embodied in particular contexts. In social theory it is best seen as a rejection of central assumptions of the modern world or of what has been described as the 'enlightenment project'. This project has had at least two core beliefs. First is the assumption that modern society will become more democratic and just because of our growing ability to rationally and objectively understand the community's best interests. Second is the assumption that scientists and social theorists hold a privileged viewpoint since they are taken to operate outside of local interests or bias. Each of these assumptions suggests the possibility of disinterested knowledge, universal truths and social progress. The late twentieth century writings of Michel Foucault and Jean Francois Lyotard called these assumption into question. Foucault's work has argued that knowledge and power are always intertwined and that the social sciences, rather than empowering human actors, have made humans into objects of inquiry and have subjected them to knowledge legitimated by the claims of science. Similarly Lyotard has argued that social theory has always imposed meaning on historical events (think of the writing of Marx) rather than providing for the understanding of the empirical significance of events. This rejection of the idea of social and intellectual progress implies that people must accept the possibility of history having no meaning or purpose, abandon the idea that we can know what is or is not true and accept that science can never create and test theories according to universal scientific principles because there is no unitary reality from which such principles can be established. We are left living in a fragmented world with multiple realities, a suspicion of science or authoritative claims and many groups involved in identity politics in order to impose their reality on others. The clearest signs of a postmodern approach to sociology can be found in social constructionism, ethnomethodology and labeling theory.

Poststructuralism/poststructural theory. Point of view that emphasizes relationships among language, subjectivity, social organization, and power.

Power. Relationship between people including the ability to control the behavior(s) of others. The potential to influence or restrict a partner's behaviors.

Power distance. According to Hofstede, the degree to which the unequal distribution of power is accepted by people in a culture (high power distance) or rejected by them (low power distance).

Prejudice. Prejudging others using positive or negative attitudes based on stereotypes rather than information about a specific individual. [IRW] To make a judgment about an individual or group of individuals on the basis of their social, physical or cultural characteristics. Such judgments are usually negative, but prejudice can also be exercised to give undue favour and advantage to members of particular groups. Prejudice is often seen as the attitudinal component of discrimination.

Primary public. The group of people an organization ultimately hopes to influence or gain approval from.

Print epoch. Third era in media history of civilization. Invention of the printing press made it possible to mass-produce written materials so that reading was no longer restricted to elite members of society.

Prototype. An organized understanding of what the defining qualities are of some category of people, events, objects, or situations.

Proxemics. Study of the use of space and of distance between individuals when they are communicating.

Public. A group of individuals tied together by a sense of common characteristics or responses.

Public affairs. That aspect of public relations dealing with the political environment of organizations.

Public communication. A multistep, multidirectional process in which messages are disseminated to a broad, and sometimes undifferentiated, audience through complex networks of active transmitters.

Public opinion. An attitudinal measure of the image a public holds concerning some person, object, or concept.

Public relations. A management function that helps define an organization's philosophy and direction by maintaining communication within a firm and with outside forces and by monitoring and helping a firm adapt to significant public opinion.

Public speaking. Presentation of a speech, usually prepared in advance, during which the speaker is the central focus of an audience's attention.

Publicity. Publication of news about an organization or person for which time or space was not purchased.

Publisher. Chief official of a newspaper who directs financial, mechanical, and administrative operations, and sometimes news and editorial operations as well.

R

Race. A classification of humans beings into different categories on the basis of their biological characteristics. There have been a variety of schemes for race classification based on physical characteristics such as skin colour, head shape, eye colour and shape, nose size and shape etc. A common classification system uses four major groups:

Caucasoid, Mongoloid, Negroid and Australoid. The term was once popular in anthropology, but has now fallen into disrepute, because the idea of racial classification has become associated with racism - the claim that there is hierarchy of races. The idea of race categories also appears to be unscientific, since humans are able to mate across all 'races' and have done so throughout history, creating an enormous variety of human genetic inheritance. In addition the defining characteristics of 'race' do not appear in all members of each so-called race, but merely occur with some degree of statistical frequency. If the defining characteristic of each 'race' does not appear in all members of each 'race' then the whole definition is clearly inadequate.

Racism. An ideology based on the idea that humans can be separated into distinct racial groups and that these groups can be ranked on a hierarchy of intelligence, ability, morality etc.

Red herring. A fallacy that uses irrelevant information to divert attention away from the real issue.

Reengineering. The fundamental rethinking and radical redesign of business processes to achieve drastic improvements in performance.

Repatriation. The process of readjustment associated with returning to one´s native culture after spending time away from it.

Resistance to change. The tendeng for employees to be unwilling to go along with organizational changes, either because of individual fears of the unknown or organizational impediments (such as structural inertia).

*Rhetorical reflecto*r. Tendency to be too flexible in expressing a position by ignoring self and telling people only what they want to hear.

Rhetorical sensitivity. Tendency to be flexible in adapting to others in communication; to base behavior on a complex network of perceived selves.

Ritual. Communicative performances that are regularly repeated in an organization and that members of an organization come to regard as familiar and routine.

Role. The typical behavior that characterizes a person in a specific social context. [A position, or status, within a social structure that is shaped by relatively precise behavioural expectations (norms). A role has been described as the active component of status. The individual, placed within a status in a social structure, performs their role in a way shaped by normative expectations. Individuals have varying ideas about normative standards and their own unique values, so role behaviour is not standardized, however radical departure from expected role behaviour will usually result in social sanctions.] A position that specifies behavioral expectations and status in relation to others.

Rules. Shared understandings about what is appropriate and inappropriate in various situations.

Rumors. Information with little basis in fact, often transmitted through informal channels.

S

Scientific management. An early approach to management and organizational behavior emphasizing the importance of designing jobs as effeciently as posible. A method of work organization where management implements a specialised division of labour and sets out detailed instructions for the performance of work. Associated with the innovative methods introduced by Frederick Taylor to separate workers from their knowledge of the work process, to divide labour so as to pay only for the specific skill required to perform a narrow function and to establish management as the controller of work and the work process.

Search engines. Programs used to find information on the World Wide Web.

Self-disclosure. Communication in which information about self normally hidden is revealed honestly and accurately to another. [Voluntary sharing of information about the self that another person is not likely to know. Disclosing personal or private information about oneself to another.

Self-fulfilling prophecy. Molding of behavior by expectations so that what was expected does indeed happen.

Self-presentation. An intentional tactic in which a person reveals certain aspects of himself or herself in order to accomplish specific goals.

Sex. The biological classification of individuals as males and females. Sociologists would note, however, that even though this is a classification based on biological differences it is a socially constructed classification. Biological and genetic quality. Sex refers to being male or female. Sex is not the same as gender.

Sexism. Actions or attitudes that discriminate against people based solely on their gender. Sexism is linked to power in that those with power are typically treated with favour and those without power are typically discriminated against. Sexism is also related to stereotypes since the discriminatory actions or attitudes are frequently based on false beliefs or over generalizations about gender and on seeing gender as relevant when it is not.

Sexist language. Language that creates sexual stereo-types or implies that one gender is superior to another.

Sign. Something that stands for or represents something else and bears a natural, nonarbitrary relationship to it. For example, dark clouds, thunder, and lightning are signs of rain.

Signal. Another category of sign. That which stands for something by virtue of a natural relationship of causality, contingency or resemblance. For example. a blinking yellow light signals drivers to slow down.

Small group. Group of fewer than 20 people who develop regular patterns of interaction and share a common purpose; members influence and are influenced by each other.

Small group communication. Communication between and among the members of a small group; communication involving several people. Exchange of information among a relatively small number of persons, usually three to thirteen, who share a common purpose, such as doing a task, solving a problem, making a decision, or sharing information.

Small talk. Casual conversation that is often impersonal and superficial, including greetings, comments about the weather, newsworthy events, or trivia.

Social loafing. The tendency for group members to exert less individual effort on an additive task as the size of the group increases.

Social relationships. Relationships in which participants interact with social roles but do not significantly depend on each other as individuals.

Source. The originator of a thought or idea subsequently transmitted to others in the communication process. Originator of a message.

Spamming. Sending unsolicited mass e-mail to members of e-mail discussion lists or Usenet newsgroups.

Speech community. A group of people who share understandings of comrnunication that are not shared by people outside of the group.

Spiral of silence theory. Explains how perceptions of public opinion can minimize social expression of minority opinion while exaggerating majority voices.

Spontaneous publicity. Publicity accompanying unplanned events.

Stakeholder analysis. A method for characterizing publics according to their interest in an issue.

Stereotypes/stereotyping. Beliefs about members of a group based on learned opinions rather than information about a specific individual. Beliefs that all members of specific groups share similar traits and are prone to behave the same way. The categorizing of events, objects, and people without regard to unique individual characteristics and qualities. [Predictive knowledge based on understandings of how members of some category can be expected to act.

Strategic plans. Long-range plans concerning a group's major goals and ways of carrying them out. These plans usually are made by top management.

Subsystem(s). Elements that make up a system. Smaller units which are part of a system's hierarchy.

Suprasystem(s). An element of systems theory that presupposes that a number of systems are interrelated to form a larger entity, the "suprasystem". Larger units which make up a system: suprasystems are composed of subsystems.

Symbol. Representation of an idea. Type of sign which is arbitrary, agreed upon, and is used to stimulate meaning. That which stands for or represents something else but bears no natural relationship to it. [Arbitrary, ambiguous, and abstract representations of other phenomena. Symbols are the basis of language, much nonverbal behavior, and human thought.

Symmetrical. A form of communcation and relationships in which power is equal between partners. Symptom. Type of sign that bears a natural relation to an object.

Synergy. A proposition underlying systems theory which maintains that the whole is greater than the sum of its parts.

Synchronous communication. Also known as simultaneous communication. Communication during which participants engage in the process at the same time, such as telephone conversation or Internet chat room.

System(s). Set of interdependent units which work together to adapt to a changing environment. An organization is one type of system. A set of objects or events grouped together by sets of relationships. Entities that function as a result of the interdependent action of its components (subsystems). A closed system is neither sensitive to its environment, nor does it make adjustments for external events (e.g., a wristwatch). An open system is sensitive to environmental influences, and may have internal mechanisms for adjustment (e.g., an organization).

T

Tactical plans. Short-range plans for accomplishing the steps that lead up to achieving an organization's goals. These plans are carried out at every level of an organization and on an everyday basis.

Target audience. In advertising, the segment of the population for whom the product or service has an appeal. The primary group an organization is trying to influence.

Taylorism. The work management principles followed by Frederick Taylor designed to transfer control of the work process to management and to achieve the greatest rate of productivity from workers through dividing labour and having work performed in a manner detailed by management.

Team. A group whose members have complementary skills and are committed to a common purpose or set of performance goals for which they hold themselves mutually accountable. A special form of group that is characterized by a close-knit relationship, members with different and complementary abilities, and a strong sense of identity.

Technological determinism. Point of view that claims that media decisively influence how individual think, fell, and act, as well as how they view collective life.

Technophile. Literally, a lover of technology. Likely to be a person who sees the positive benefits deriving from technology and advocating increased use of technology as a solution to economic, social and political problems within the society.

Technophobia. Literally, the fear of technology.

Theory. An account of what something is and/or how it works and/ or what it produces or causes to happen and/or what should be the case. Theories are points of view, human constructions. A tentative but usually systematic explanation for a problematic situation; an educated guess open to change. An explanation or belief about how something works.

Transaction. A relationship in which each party simultaneously defines and is defined by the other.

Tribal epoch. First era in media history of civilization. During the tribal epoch, the oral tradition reigned and face-to-face talking and listening were primary forms of communication.

Trust. Faith in the behavior of another. Promotes confidence in risk taking. The belief among employees that they will be treated fairly by their organization and, more specifically, by their immediate supervisor.

Trustworthiness. Audience's perception of a speaker's reliability and dependability.

Turn-taking. Behavior which exhibits the exchange of source and receiver roles during conversation.

Two-step flow. Theory which asserts that information from media is processed first by opinion leaders who then pass it along via interpersonal channels.

U

Uncertainty avoidance. According to Hofstede, the degree to which people in a culture feel threatened by, and attempt to avoid, ambiguous situations.

Uncertainty reduction theory. Point of view that claims uncertainty motivates communication and certainty reduces the motivation to communicate.

Uncontrolled media. Those media whose actions are not under the public relations practitioner's control, such as community newspapers and radio.

Upward communication. Communication from lower members of the organizational hierarchy (subordinates) to members higher in the organization (i.e., managers, vice-presidents).

Uses and gratifications approach. Explains how mass media audiences make choice to use media content for their own gratifications and their own purposes.

V

Value(s). Relatively general cultural prescriptions of what is right, moral and desirable. Values provide the broad foundations for specific normative regulation of social interaction. A general, relatively long-lasting ideal that guides behavior.

Verbal communication. One of two major communication code systems; associated with our spoken and written language; typically labeled as "language". The transmission of messages using words, either written or spoken.

Vertical chain of communication. Communication between members of different levels of organizational hierarchy; i.e., between managers and subordinates.

W

Whistle-blowing. Insiders telling the media what they know about improper practices by others, usually in the same company, with the hope of improving the situation. Calling attention to actions or practices that are inconsistent with established organizational norms or policies.

Work team. A group of people who are responsible for a whole work process or a segment of the process that delivers a product or service to an internal or external customer.

X

Xenophobia. An individual's irrational and obsessive hatred of people perceived as different and foreign. Related to the concepts of racism and ethnocentrism. All of these can be overcome by the study of the social sciences and coming to appreciate the ideas of culture and social structure as tools for understanding ourselves and others.

□□□

References

Anderson, R. & V. Ross 1994. Questions of Communication. A Practical Introduction to Theory. New York, NY: St. Martin´s.

Baskin, O. & C. Aronoff 1992. Public Relations. The Profession and the Practice. Wm. C. Brown.

Dominick, J. 1987. The Dynamics of Mass Communication. 2nd Edition. New York, NY: Random House.

Drislane, R. & G. Parkinson 2001. Online Dictionary of Social Sciences. Athabasca University, Canada. Online: http://datadump.icaap.org/cgi-bin/glossary/SocialDict, quoted February 6, 2001.

Greenberg, J. & R. Baron 1995. Behavior in Organizations. Undestanding and managing the Human Side of Work. 6th Edition. London: Prentice-Hall.

Infante, D., A. Rancer & D. Womack 1997. Building Communication Theory. 3rd Edition. Prospect Heights, IL: Waveland Press.

Seiler, W. & M. Beall 1999. Communication. Making Connections. Boston, MA: Allyn and Bacon.

Stacks, D., M. Hickson & S. Hill 1991. Introduction to Communication Theory. London: Holt, Rinehart & Winston.

www.ugc.com

Wilcox, D., P. Ault, W. Agee & G. Cameron 2000. Public Relations: Strategies and Tactics. 6th Edition. New York, NY: Longman

Witmer, D. 2000. Spinning the Web: A Handbook for Public Relations on the Internet. New York: Longman.

Wood, J. 1997. Communication Theories in Action. Belmont: Wordsworth.

Wood, J. 1995. Relational Communication. Continuity and Change in Personal Relationships. London: Wordsworth.

❑❑❑